Ethnobotany for Foresters

NIPA® GENX ELECTRONIC RESOURCES & SOLUTIONS P. LTD.
New Delhi-110 034

Ethnobotany for Foresters

Kaushal Singh
Assistant Professor
Department of Silviculture and Agroforestry, College of Forestry
Banda University of Agriculture and Technology
Banda-210001, Uttar Pradesh

Vinita Bisht
Assistant Professor
Department of Silviculture and Agroforestry, College of Forestry
Banda University of Agriculture and Technology
Banda-210001, Uttar Pradesh

Aishwarya Maheshwari
Assistant Professor
Wildlife Sciences
Banda University of Agriculture and Technology
Banda-210001, Uttar Pradesh

Manish Kumar Singh
Scientist-D
Rain Forest Research Institute
Jorhat-785 010, Assam

NIPA® GENX ELECTRONIC RESOURCES & SOLUTIONS P. LTD.
New Delhi-110 034

NIPA® GENX ELECTRONIC RESOURCES & SOLUTIONS P. LTD.

101,103, Vikas Surya Plaza, CU Block
L.S.C. Market, Pitam Pura, New Delhi-110 034
Ph : +91 11 27341616, 27341717, 27341718
E-mail: newindiapublishingagency@gmail.com
Web: www.nipabooks.com

For customer assistance, please contact
Phone: + 91-11-27 34 17 17
Fax: + 91-11- 27 34 16 16
E-Mail: feedbacks@nipabooks.com

ISBN: 978-81-19103-82-9

Composed and Designed by NIPA®.

डॉ. यू. एस. गौतम
कुलपति
Dr. U.S. Gautam
Vice Chancellor

बाँदा कृषि एवं प्रौद्योगिक विश्वविद्यालय
बाँदा – 210001, उ.प्र., भारत
BANDA UNIVERSITY OF AGRICULTURE & TECHNOLOGY
BANDA-210001, U.P., INDIA

Foreword

Every person has association and some knowledge of plants. In addition to provide Oxygen, plants have been used for various purposes. Despite the fact that medicinal plants have an affluent narration of utilization in every culture but lacking substantial information on when or where plants first began to be used in healing of various ailments. Since, existence of human on earth, they have familiarized themselves with plants and have used them in a diverse ways throughout the ages. Primitive humans began to differentiate those plants species suitable for nutritional purpose from those with a definitive therapeutic action. This information has grown among humans and many plants have come to be used as drugs. Later on this traditional knowledge of indigenous communities about the surrounding plant diversity and the study of how the people of a particular culture and region make use of indigenous plants arose as branch of science known as ethnobotany. India is one of the most religiously and ethnically diverse nations in the world with some of the most deeply religious societies and cultures. With over 84.4 million, India is having the largest population of the tribal assemblage in the world. Also in India, there are approx 45,000 species of wild plant out of which 9,500 species are ethnobotanically important species. Of these 7,500 species are in medicinal use for indigenous health practices. Different societies use plants according to their own beliefs, knowledge and previous experiences. Their knowledge about the use of the plants is usually not known to other societies or to scientists. These hidden areas need to be explored for our future generation.

Now a days, several academic institutions have introduced forestry subject in their undergraduate (UG) and postgraduate (PG) courses and ethnobotany is being taught at UG and PG levels either as a separate subject or as a part of the regular curricula of agriculture, forestry, botany, ecology and other related programs. There are continuous contributions towards documenting ethnobotanical practices and this book is an effort to make it more meaningful.

I congratulate my colleagues, Drs. Kaushal, Manish, Vinita and Aishwarya for compiling this book and may be used for academic, professional and associated various purposes. I hope that this book will provide much needed literature on ethnobotany.

U.S. Gautam
Vice Chancellor

Acknowledgements

The success and final outcome of this book required a lot of guidance and assistance from many people and we are extremely privileged to have got this all along the completion of our book. All that we have done is only due to such supervision and assistance and we would not forget to thank them.

We express our sincere gratitude to Dr. S.K. Jain, pioneer ethnobotanist of India, former Director, BSI Kolkata for sharing his valuable experience in ethnobotany during a training programme at NBRI, Lucknow and encourging to accomplish this task.

We thank the Hon'ble Vice Chancelor, BUAT, Banda for constant encouragement, support and time to time guidance which helped us to complete this book.

We also thank Mr. Dinesh Meena, Scientist, RFRI, Jorhat, for providing us distribution map of Major tribes of India and providing us photograpshs of some of the tribes North eastern India. We are extremely thankful to him for providing such a nice support, without which the book would not have been completed.

We owe our deep gratitude to Dr. Praveen Verma, Scientist, Shri N.K. Limba, AFRI, Jodhpur and Dr. Manoj Kumar, FRI, Dehradun for sharing his experience, knowledge on ethnomedicinal plants and providing us necessary information during writing this book.

We are thankful to and fortunate enough to get constant encouragement, support and guidance from all our friends and colleagues and our family members which helped us in successfully completing our book. Also, we would like to extend our sincere esteems to all the informants who shared with us their experience and knowledge while writing Ethnobotany for Foresters book.

Authors

Contents

1

Introduction

John William Harshberger codified ethnobotany as 'the study of plants used by primitive and aboriginal people 'in a popular lecture on 4 December 1895. Later it was published in the Botanical Gazette in 1896 and now he is regarded as "Father of Ethnobotany". Subsequently, Walter Fewkes (1896) adopted the word ethnobotany after Harshberger for the title of a research paper. Another term *'Arboriginal Botany'* coined by Steven Powers (1873- 1875) was in use for about two decades before Harsh Berger. In brief, "Ethnobotany is the study of interrelations between primitive human cultures and the plants and animals in its environment, which is regarded as the branch of ethnobiology". There are some other definitions of ethnobotany by various authors:

Jones (1941) defined as "Ethnobotany is the study of interrelationship between man and vegetation".

Faulks (1958) considered it as "The total relationship between man and vegetation" in his book entitled "An Introduction to Ethnobotany" It was the first book specially published on ethnobotany. Schultes (1962) explained as "Ethnobotany is the study of relationships which exist between people of a primitive society and their plant environment". Gadgil and Vartak (1976) considered it as "Branch of economic botany which deals with the role of plants in life and culture of arboriginals and tribal people". Alcorn (1984) states "It is the study of contextualized plant use". Wickens (1990) defined ethnobotany as "The study of useful plants prior to their commercial exploitation and eventual domestication". According to Ford 1994, ethnobotany is concerned with a wide range of interest of plants in cultural and ecological context as men in his edited book "The Nature and Status of Ethnobotany". According to Martin 1995, ethnobotany is the part of ethnoecology, which concerns plants. Turner 1996 was given the appropriate definition "the science of people's interaction with plants". In context of Indian ethnobotany the first book published by Jain, S.K., 1981 entitled "Glimpses of Indian Ethnobotany" and Dr. Jain (2001) defined ethnobotany as "It deals with the study of total natural and traditional interrelationship between man and plants and his environment".

In broad terms, ethnobotany is the study of the interrelationship between plants and people. The two major parts of ethnobotany are encapsulated in the word itself **ethno**: 'the study of people' and **botany**, 'the study of plants'. However, the field is limited on both sides. On the botanical sides of the field, few ethnobotanical studies are concerned with plants that are not related to the people. On the ethno side, most studies are concerned with the ways indigenous people's use and view plants. And those uses and views can provide deep insights into the human conditions (Balick and Cox 1996).

Ethnobotany represents collective research in botany, archaeology and allied subjects. Ethnobotany requires skillful observations and recording of ecological information through rigorous fieldwork and gathering first hand data. Traditionally, communities have been using various flora for numerous purposes. Scientific documentation of such knowledge is crucial and requires specialized botanical and taxonomic skills. Plants have been associated with human ever since ancient times and their usage may be classified in two groups *Cultural* and *Material*

1. Cultural: It includes faith in God or plants taboos, worship and folklores.
2. Material: It includes the material and services provided by the flora such as food, medicine, house construction material, agricultural tools and other domestic uses, plants in painting and carving.

Indigenous communities may have their traditional conservation values and practices but current unsustainable-use (such as intensive agriculture, excessive livestock grazing) have led to degradation of the native flora.

Historical review of ethnobotany

Exact or first usage of plants in human civilization is not known. However, the evidence of unique collection of plant products exhibited in World's Fair at Chicago in 1893 might be one of the early records in ethnobotany. This collection might have inspired W. Harshberger to explore a new area of study and he wrote an article in the Botanical Gazette (1896) 'The study of people-plant relationships ethnobotany'. In his opinion, ethno-botany can expose the cultural position of different tribes whose use the plants for shelter, food, clothing and ailment. These plants are known as 'ethnobotanicals'. He explained about indigenous people of southwest U.S. 'roving people, travelling from place to place in search of game and settling only long enough to cultivate some corn, beans and pumpkins to break the monotony of a too strict animal diet. Where they did not pursue agriculture, they subsisted on the seeds of wild grasses and herbs. The cliff dwelling people, apparently driven to the mountain fastness, had practically left the hunter stage and begun to enter the agricultural stage' (Harshberger, 1896). Richard Ford in 1994 published an elegant "tree ring" scheme to represent the evolution of the ethnobotany as a discipline. The cultural assumption of evolution proposed by Harshberger was not accepted until a long time. Richard Evans further contributed in ethnobotany. He compiled research work and photos of Amazonian traditional healers. Nolan (2002) proposed the cognitive magnitude of ethnobotany in which, he established interrelations between language and memory in human society. Deur and Turner (2005) proposed a Traditional Ecological Knowledge (TEK) system, which include diverse traditional land resource management methods of important aspects of ethnobotany. Now ethnobotany became popular at national and international levels and the first International Congress of Ethnobotany was held in Istanbul in 2005. In India, it seems that initially, Dr. S. K. Jain referred the term ethnobotany at the Central Drug Research Institute (CDRI), Lucknow during his tenure at the Economic Botanist in Botanical Survey of India (Personal communication, 14th December, 2018) during Lifetime Achievement Award by BSI at National Botanical Research Institute, Lucknow. A post which Dr. Sudhanshu Kumar Jain as the first to hold this become the main ground for convincing the Director of R.C. of Central Drug Research Institute at Lucknow to change the name of Botany Division to Ethnobotany. The first edited book by Dr. S. K. Jain in 1981 on the subject was a compilation of two dozen papers on various ethnic groups belonging to different regions. This work triggered interest in many botanists to explore the ethnobotany.

Ethnobotany is an important subject. It plays an important role in production of food and medicines indigenously. It also increases mutual understanding between bio-diversity and human culture.

Dimensions of Ethnobotany

As proposed by Dr. Jain, there are following dimensions of ethnobotany:

i. Plants referred in epics of various faiths.

ii. Plants related with life or deities of different faiths such as Hinduism, Islam, Christianity, Buddhism and Jainism.

iii. Plants related with planets and astronomical phenomena.

iv. Dynamism in ethnobotany: ethnobotany of any region or ethnic group is never static and these should be targeted for study of changes after 25-30 years.

v. Credibility of folk claims: There should be some quantitative, qualitative, analytical and statistical models.

vi. Use of plant parts like leaf, root, flower, fruit etc. for a disease in different regions or among several ethnic groups.

vii. Group statement by local medicine man about the effectiveness over decades of his experience.

viii. Urban ethnobotany: it is recent concept by some able ethnobotanists who perhaps did not have access to areas rich in natural vegetation and un-civilized ethnic groups.

Scope and objectives

Since the origin of human civilization, people have been used to plants for providing them food, shelter, medicines, and the materials for construction as well as the manufacturing of the crafts and tools/equipment and many other products such as fuel, paints, poison, etc. Plants often used for ritual purposes. Continuous ethnobotanical studies providing meaningful insights in the plants usage at a much wider spectrum for the benefits of the society.

The definition as well as the scope of ethnobotany has remained impressive even by the narrowest definition of the discipline. This is an interdisciplinary science and undertakes a research on the relationship between people and plants in the areas of linguistics, education, healing, nutrition, archaeology, paleology, resource tenure and management, livelihood, etc. Ethnobotany may therefore serve as a gateway to many disciplines. The following disciplines are often included within the study of ethnobotany:

Discipline Relevance to Ethnobotany

Botany	Study of plants
Anthropology	The study of how different cultures use plants
Ecology	How human interactions with plants and ecosystems affect ecology
Medicine	Study of medicinal uses of plants
Chemistry	Study of composition of plants, especially medicinal plants
Agriculture	Study of plant management and human domestication, especially traditional agriculture system
Horticulture	Study of management of useful plants (fruits, vegetables, ornamentals) in home garden or orchard
Forestry	Study of human management of forest and forest trees
Agroforestry	Study of land management for the simultaneous production of food, crops and trees
Archaeology	How ancient cultures used plants

Economics	Study of economic uses of plants
Religious studies	Ritual uses of plants by different cultures and religions
Linguistics	Study of linguistic terminology for plants and plant parts by people of different language groups
Systematics	Study of folk taxonomy, how different people classify plants.

As we have seen, ethnobotany is a multidisciplinary science and its scope is not confined to one area but it covers a broad range of study areas, which are interconnected to each other in one sense or the other. Therefore, there is a great opportunity to explore the ethnobotanical approach towards the modern civilization and giving them a firm task, which should include:

- Conservation of plant species- including varieties of crops and other forms of biological diversity.
- Botanical inventories and assessment of the conservation status of the species.
- Sustainability in supplies of wild plant resources.
- Enhanced food security, nutrition and healthcare.
- Preservation, recovery and diffusion of local botanical knowledge and wisdom.
- Reinforcement of ethnic and national identity.
- Identification and development of new economic products from plants, for instance food, crafts, herbal formulations, horticultural plants etc.

Terms employed in ethnobotany

Abortifacient: A prescription believed to induces the premature abortion of the fetus.

Adaptogens: Herbs that help in acclimatizing stress level by supporting the adrenal glands and endocrine system. Eg. ginseng root, nettle leaf, sarsaparilla, licorice root and ashwagandha.

Alterative: These herbs help in elimination of metabolic toxins and improve lymphatic circulation, boost immunity and help in clearing chronic conditions of skin. Such prescriptions are also believed in healing sores, boils, tumors, cancers, reduces fevers, detoxifies the liver, kills parasites and worms, helps in the treatment of infectious, contagious diseases and epidemics, flu, acne herpes, and venereal disease. Eg. ginseng, aloe, sandalwood, red clover, burdock, bayberry, black pepper, cinnamon, myrrh and safflower.

Analgesic or anodynes: These herbs reduce pain (e.g., digestive, circulatory, respiratory, nervous system, nerve, muscle, tooth pain, nervous digestion, headaches). Eg. Camphor, chamomile, cinnamon, cloves, echinacea, lavender flower, feverfew herb, cabbage leaves, wintergreen leaf, passionflower herb and flower.

Anaphrodisiac: A prescription believed to increase sexual desire.

Anesthetics: A prescription used before surgical operations. Eg. ashok, calamus, gudmar and jatamanshi.

Anodyne: A prescription helps in relieving from pain and reduces the sensitivity of the nerves. Eg. Ashok, barberry, cedar and ginger.

Antacid: A herb or formulation that neutralize the access stomach acidity. For example, fruit pulp of *Cassia fistula*, levaes of *Cinnamomum zeylanicum* etc.

Anthelmintic: An agent that destroys and expels worms from the intestines and works as vermifuge.

Anthelmintic: Herbs that destroys and dispels worms, parasites, fungus, yeast such as vermicide and vermifuge. Eg. goldenseal, wormseed, wormwood, ajwan, cayenne, peppers and pumpkin seeds.

Antibiotic: A prescription that inhibits the growth of germs, bacteria, and harmful microbes. Eg. Turmeric and Echinacea.

Antidiabetic: A prescription that helps with diabetes control, meaning level of glucose (sugar) in the blood. Eg. Amalaki, blackberry, fenugreek, gudmar, senna and shilajit.

Antidiarrhea: A prescription that uses against diarrhea. Eg. Blackberry, comfrey, gentian, red raspberry, and yellow dock, black pepper and ginger.

Antiemetic: Herbs helps in preventing nausea and vomiting. Eg. Cloves, coriander, ginger, raspberry.

Antiepileptic: Herb that helps in epilepsy.

Antilithic: A prescription believed in preventing the formation of stones in the kidneys and bladder.

Antiperiodic: A prescription prevents the periodic recurrence of attacks of a disease such as in malaria. Eg. Barberry, chirayata, guduchi, kutaj and vacha.

Antipyretic: A prescription helps in reduceing fever through sweating. Eg. Amalaki, black pepper, brihati, nirgundi, safflower, sandalwood.

Antiseptic: A prescription prevents putrefaction through checking the growth and development of microorganisms. Eg. Aloe, Chitrak, gokshura, gudmar, sandalwood and turmeric.

Antispasmodic: Herbs prescribed in relieving from involuntary muscle spasm by strengthening the nervous system. Eg. Camomile, ashwagandha, basil, calamus, guggul, licorice, myrrh, sage, gotu kola, jatamanshi, peppermint, sandalwood and spearmint.

Antisyphilitic: An antiluetic prescription cures syphilis. Eg. Black pepper, cedar, guduchi, guggul.

Aperient: A mild or gentle laxative. Eg. Rhubarb.

Aphrodisiac: A prescription believed to increase sexual desire.

Aphrodisiac: A prescription believed to increase sexual desire. Two types of tonics are prescribed: 1. Develop tissue substance. Stimulants: increase the functioning of the reproductive organs. Eg. Angelica, ahwagandha, asparagus, fenugreek, ginseng, gokshura, hibiscus, kapikachu seeds, pippali, rose, saffron and shatavari. 2. Nutritive tonics such as aghwagandha, bala, ghee, licorice, marshmallow, sesame seeds and Shatavari increase semen and breast milk.

Appetizer: Herbs that stimulate the appetite. Eg. Cardamom, coriander.

Aromatic: Herb with a pleasant fragrance. Eg. Cardamom, cinnamon, cloves, fennel, peppermint and turmeric.

Astringent: A prescription helps in arresting the discharge of blood through contraction of skin, blood vessels and other tissues. Eg. amalaki, arjuna, ashok, cinnamon, jasmine, sandalwood and yarrow.

Carminative: A herb or arrangement proposed to either forestall development of gas in the gastrointestinal tract or encourage the ejection of said gas.

Carminative: Herb that helps to prevent gas formation in the intestines and also assists in expelling it. It is also believed in improving weak digestion from anxiety, nervousness, or depression. Eg. Chamomile, chrysanthemum, coriander, fennel, lime, peppermint, and spearmint, ajwan, basil, calamus, cardamom, cinnamon, ginger and turmeric.

Demulcent: A prescription believed in relieving irritation of inflamed mucous membranes i.e. stomach and urinary bladder lining. Eg. Barley, licorice; linseed, olive and almond oils.

Diaphoretic: These herbs are believed to use for various purposes such as eases muscle pressure, joints pain, fiery skin conditions, anti-diarrhea, smooth functioning of kidneys, liver, urinary, and gall bladder and urinary infections. Eg. basil, ajwan, cardamom, parsley, cinnamon, eucalyptus, ginger, juniper berries, asparagus, barley, burdock, chamomile, chrysanthemum, coriander, dandelion, fennel, marshmallow and spearmint.

Digestives: These herbs are believed to improve gastrointensinal digestion of food. Eg. Coriander, cumin, rock salt, and turmeric.

Diphoretic: Drug that increases sweating and helps in temperature regulation.

Discutient: Herbs have discussive properties to help in curing tumors.

Disinfectant: These herbs are believed to kill harmful germs and noxious properties of fermentation. Eg. Apamarga, arka, gudachi, katuka, sandalwood.

Diuretic: A prescription that helps production and secretion of urine. Eg. Parsley, apamarga, ashwagandha, barberry, cardamom, cinnamon, ginger, gotu kola, gokshura, gudachi, licorice and sandalwood.

Emollient: A prescription that believed soften skin. Eg. oils, honey, bread or bran poultice, carrots, turnips.

Errhine: A prescription that helps in nasal secretion. Eg. black pepper, ginger, amalaki, apamarga and arka.

Exanthematous: These herbs are believed to improve skin eruptions such as measles, scarlet fever, etc.

Exhilarant: These herbs are believed to cheer the mind.

Expectorant: A prescription that helps in curing cough through thinning and ejection of mucus from the lungs, bronchi and trachea. Eg. ginger, licorice, calamus, cardamom, cinnamon, cloves, sage, eucalyptus, thyme and wild cherry.

Expectorant: A prescription that raises bodily fluid and other material from the lungs, bronchi, and trachea.

Febrifuge: Similar to antipyretic and refrigerant, a prescription that helps in reducing body temperature.

Febriguge: A prescription helps in reducing body temperature and to relieve from fever.

Galactogogue: Herbs that stimulate breast milk secretion. Eg. Cumin, fennel, musta and pippali.

Hemostatic: A prescription that helps in clotting and purification of blood. Eg. durba, goldenseal, red raspberry, turmeric, cinnamon, ginger.

Laxative: And also known as purgatives are substances that help in loosen stools and increase viscus movements and also accustomed in treating constipation.

Narcotic: A prescription stimulates sleep and reduces pain.

Nauseant: Herbs causes nausea and vomiting.

Nervine: A prescription that calms and soothes the nerves and reduces tension and anxiety. Eg. ashwagandha, bala and gudmar.

Opthalmic: A remedy for diseases of the eye.

Parturient: A substance that induces and promotes labor.

Parturifacient: A prescription induces child-birth. Eg. Barley, pippali.

Purgative: A prescription promotes the vigorous evacuation of the bowels and usually used to relieve severe constipation. Eg. Aloe, Epsom salt, licorice, rhubarb, safflower and senna.

Refrigerant: A cooling prescription relieves fever and thirst. Eg. Aloe, coriander, ginger, hibiscus, orange, lemon, licorice, sandalwood.

Rubefacient: A prescription that reddens the skin by increasing the circulation when rubbed on the surface. Eg. black pepper, cayenne, ginger, licorice, mustard.

Sedative: A prescription relieves excitement and conducive to sleep.

Stomachic: Herbs that stimulate digestion and improve the appetite. Eg. Amalaki, bilwa, black pepper, cardamom, cedar, chitrak, cumin, ginger, licorice, turmeric.

Tonic: A prescription helps in restoring and strengthening the entire system. A general tonic would be one that braces up the whole system. For example Aloe, barberry, chirayata, guduchi, katuka, gentian and golden seal.

2

Ethnobotany and Subdisciplines

In the starting of twentieth century, the ethnobotany progressed from the untested compilation of knowledge to a bigger method and abstract reorientation. Nowadays, the ethnobotany practices need a spread of skills such as:

- Training on the identification and preservation of plant specimens
- Training on linguistic and sociology to understand local language and their family structure
- Training on the psychology and cultural or traditional about plants
- Training on pharmecology and medicobotany

Ethnobotanists interact in a very broad array of analysis queries and practices, that don't lend themselves to straightforward categorization. However, the subsequent headings arrange to describe a number of the key areas of recent ethnobotanical study.

Archaeoethnobotany

It is the study of the plant materials from archaeologic sites for studies on mass movement of human traditions, spreading of his experiences and cultivation of various crops. It's closely connected to ethnobotany, because it is tough to know the status of soil that is recently formed without knowing the history of rock or its weathering that always involves prehistoric environmental interventions. There are an ample numbers of examples from archaeologic remains for example: twigs, charcoal, hard-coated seeds etc, those were found during the excavation of various sites in man made structures. These biolocal remains reveals certain information about new source of ethnomedicine, food, fodder, fuel, dye etc. These remains also helpful in tracing the origin of human, migration of culture and domestication of many species and verieties of agriculture crops.

Histroy of crop domestication reveals about crop domestication. Corn (*Zea maiz*), cotton (*Gossypium hirsutum*), sweet potato (*Ipomoea batatas*, bel pepper (*Capsicum annum*), Papaya (*Carica papaya*), Guava (*Psidium guajaba*), cashew nut (*Annacardium occidentale*), potato (*Solanum *tuberosum*), tomato (*Lycopersicun esculentum*), peanut (*Arachis hypogaea*), rubber tree (*Hevea brasiliensis*) etc. are domesticated in America. Durum or pasta or macroni wheat or hard wheat (*Triticum durum*) tetraploid (4x) and the common or bread wheat (*Triticum aestivum*) hexapolyploid (6x), Einkorn wheat (*Triticum monococcum*) diploid (2x) etc. were domesticated in Asia. Pearl millet (*Pennisetum glaucum*), Jawar (*Sorghum bicolor*), til (*Sesamum indicum*) were domesticated in West Africa. Sugarcane (*Saccharum officinarum*), mango, tamarind, banana etc were domesticated in India. Whereas the rice (*Oryza sativa*), soyabean, apricot, litchi orange etc were domesticated in China.

Ethnoecology

Ethnoecology is the scientific study of the manner within which different teams of the folks in several locations perceive ecosystemsaroundthem; the environments in which they live; and their relationship with these. It seeks valid, reliable understanding of how we have a tendancy to as humans have interacted with the surroundings and the way these complex relationships are sustained over time. The term ethnoecology comprises two terms ethno and ecology. Here ethno indicates a localized study of people, and other is ecology that signifies people's understanding and knowledge of around them. Some definitions of ethnoecology are as follows:

"Ethnoecology is science of how people understand the relationship between human, animals, plants and physical elements of local environment" (Davison-Hunt, 2000).

"Ethnoecology encompasses all the studies which describes local people interaction with natural environment including sub discipline such as ethnobiology, ethnobotany, ethnoentomology and ethnozoology" (Martin, 1995).

Thus, ethnoecology may be a multidisciplinary field that integrates technique from biology, social science, ethnology, linguistics, economy and associated fields. The work of an ethnoecologist is not confined up to the forest only the trees out side the forest which are growing nearby the habitations or along the agriculture fields, which are affected by humain activities or culture, also included in his/her study. Ethnoecologist conduct surveys in villages of forest fringe area and note down the various information related to ecological parameters and human traditions in that village or region.

Ethnomedicine

This is the study of traditional medicines-not only those with relevant written sources like Ayurveda but also the knowledge and practices, which are passing orally form one generation to another generation from thousands of years.

While the focal point of ethnomedical studies is regularly the indigenous recognition and utilization of customary medications, another upgrade for this sort of research is sedate revelation and improvement. Real pharmaceuticals like quinine, morphine, and atropine have been extracted from cincona, opium, and belladonna, individually. Ethnomedical examinations in this century have prompted the improvement and upgrade of significant medications, for example, Colchicine used in treatment of tumor and gout, morphine used as analgesic and cafeine used as central nervous system (CNS) stimulant. In the scientific space, ethnomedical studies are usually characterised by a powerful social science approach, or by study medicine approaches in invention of new medicines. The main focus of social science studies is that the perception of use of the standard medicines, whereas medicine approaches typically target the discovering therapeutic new compounds for the treatment of tumor and Dengu.

Ethnogynaecology

Ethnogynaecology is the investigation, which manages different infections among women in inborn social orders, identified with sterility, origination, fetus removal, and so forth, and the utilization of abortifacients. The medicines got from the plants to treat certain gynecological issues have been accounted for to be helpful and safe. *Abrus precatorius*, *Adhetoda vasica, Pedalium murex*, *Desmodium heterocarpum, Ocimum sanctum, Smilax zeylanica* are a few instances of the plants, which have been demonstrated useful in treating gynecological diorders. *Withania somnifera, Lawsonia innermis* were utilized as abortifacient by people of various tribes since ages as a traditional method.

Ethnomusicology

Ethnomusicology comprises two words Ethno means people and musicology means cultural aspect of music. Ethnomusicology is the study of music and dance of the people of various culture and traditions. It includes the various types of dances performed by various tribes separately by mens and womens on different occasions and festivals with the combination of various types of songs and bands. Like people of some tribe, dance in-group and some dance solitary during their respective festivals in their region.The song *Jaccha* is sing during the birth of a new baby in family with music of barrel in Rajasthan whereas in UP and Madhya Pradesh these are known as the *Sowar or Sohar.* Saranagi, Ektara, Morchang, Kamayacha and Rawanhatta are some of the musical instruments of Rajasthan. The *Daf* and *Chang* musical instruments play during the Holi festival.

The Manganiars and Langa tribes of Rajasthan sing specific *ragas* in specifc season at specific time of the day on different festivals like wedding and birth ceremoney. The songs like Jhorva, Papiha and Kuranjaare sing by the womens of Jaisalmer in western Rajasthan wheras Hichki song is sing in Mewat region in memoary of his husband that is far from her. For the same purpose, the folk song Suvatia is sing by the womens of Bhil tribe in Mewar region. The song Peepli is sing in Bikaner and Shekhawati region by the womens during rainy season. *Malhar* and *Hindolya* songs sing in Shravan month whereas *Ghoomar* sing with dance on *Gangaur* and *Teej* festival in northwestern and central India. The song, which is sing in Falgun month, known as *Faag* as well as *Rasia* in central and northwestern India. *Sithane* or *Garee*s are sing by the womens of bride for the father and other relatives of bridegroom during marriage ceremony for enjoyment. *Dhudla* song is sing in Marwar region Dhudlaparva. *Iduri* song is sing by the womens of Rajasthan during going to carry the water. *Langurias* songs are used to sing in worship of mata Jwaladevi, Veshnodevi and Karolidevi. The song *Oluen* is sing during the departer of daughter from his father's house after marriage. When knewly married bride comes first time at his husband house a singning programme is organized that is known as *Dadra*. The song Lab mostly sing in a group during harvesting of crop in north- western India.

Ethnomycology

This investigation of the authentic uses and sociological effect of growths and which can be considered as a sub-field of ethnobotany. In spite of the fact that in this hypothesis the term incorporates growths utilized for such purposes as timber, medication and nourishment, it is now and again utilized with regards to the investigation of psychoactive mushrooms, for example, *Claviceps* spp. and fly amanita are the common species of mushrooms, which are collected from lower hills of Jammu & Kashmir, Himachal Pradesh and Uttarakhand.

One of the significant discussions among the ethnomycologists is Wasson's hypothesis that the Soma referenced in the Rigveda of the India was the fly amanita mushrooms. The fungus *Cleorotus* sp. is an edible mushroom found naturally on medium heights of Himalaya hills in Jammu & Kashmir, Himachal Pradesh and Uttarakhand. People sell this mushroom in local markets after drying in powdery form. *Morchella esculenta* commonly known as sponge mushroom or yellow morale, in Uttarakhand gucchi and in Himachal Pradesh Chunchuru is an edible fungus, which grow on high altitude in coniferous forests of Himachal Pradesh and Uttarakhand. *Macrolepiota procera* commonly known as parsol mushroom grow in on the hills of lower to middle altidue of Jammu & Kashmir, Himachal Pradesh and Uttarakhand. This is also edible which, is used by the grazers when they stay in grasslands with their cattles durin rainy season. *Cordyceps sinensis* locally known as keeda jari, which is found on high altitudes (Alpine) of Uttarakhand during early rainy season. It is a medicinally important parsite type of fungus which parasitoid the dead larvae of a lepdopteran. It is always in high demand in Chinese-Tibetan traditional medicine.

Ethnonarcotics

Ethnonarcotics is the investigation of the utilization of opiates, snuffs, psychedelic drugs, and so forth, in crude human social orders. Narcotic medications, likewise known sedatives or opiates, which have been utilized since the start of written history. Opium is gotten from a white fluid or latex secreated by the poppy plant, first time reported from Asia. It has been utilized restoratively in antiquated societies to incite quiet and to soothe torment.The tobacco (*Nicotiana tabacum*) plant was first brought into India by the Portuguese explorers about the year 1605. Tobacco and bhang are the main source of tradition narcotics since long time back. Bhang (*Canabis sativa*) used in three types of preparations like bhang, the green leaves of plant is collected dried and mixed with water and milk; Charas, is a pure latex or resin of the plant which is normally smoked in cigrettes and bidis or eaten with spices third one is ganja, it is dried top twigs of female plants which is rich in resin content and smoked with tobacco. Tobacco is smoke with various types of water pipes known as *hukka*, *chilam*, *bidi etc.*

Ethnopharmacology

A study that connects ethnic people or various cultural groups with their physical propensities and technique in making and utilizing medicines is called ethnopharmecology. In other words, it is a Study that deals with medical uses of plants and their bioactive compounds by various cultural groups. Indian Materia Medica is an important source of information on ethnic folklore practices and preparations of traditional medicines. Historically, all the medicines have been originated from the ethnopharmacology. Traditional medicinal system over the year has proved to be valuable guide for screening of drugs in present and future. There are many examples of modern modern drugs that have been discovered from traditional medicines like ephedrine from *Ephedra* sp., atropine from *Atropa belladonna*, digitoxine from *Digitalis purpurea*, reserpine from *Rauvolfia serpentina*, quinine from *Cincona officinalis*, vincrstine and vinblastine from *Vinca rosea* etc.

Ethnotoxicology

Ethnotoxicology is the study of various poisonous plants used by tribal people as fish, arrow poison etc. Every plant contains some of phyto-chemicals which were exploited by many ethnic groups for their safty or in the form of arrow wars in ancient time. For example *Aloe barbedensis, Datura stramonium, Euphorbia pulcherrima, Jatropha curcas, Lantana camara, Ricinus communis* etc., are some examples of toxic plants, which have been used by ancient people from time immemorial.

Ethnocosmetics

It is the study of beauty care products which were used by womens since past for improve their fairness. Thes beuty products may be derieved from plant or animals or may be various types of practices in which variuos plants were used. Hina or Mehandi has for quitesome time been utilized in India for shading hand palms and feet during festivals or marriage ceremonies.Turmeric (*Curcuma longa*) Chandan (*Santalum album*) and rose (*Rosa domescena*) are the very common plants which have been reported to be utilized by womens since time immemoral for improve their beauty by nourishing the skin and utilizing the aroma of rose for bathing and as a face-wash or as esteam therapy. Clay soil (Multani mitti) was reported to be used from long time for removing the dead cells from skin and making the skin and hair more smoother or shiny. There are ample of evidence of highly advance cosmetics uses in ancient India. The oil of Til seeds (*Sesamum indicum*) and ghee used for making various types of *Lepas* for increasing glow in skin. A herbal *Lepas* or paste that is known as *Ubtan* used by womens since long time for glowing skin. The ingradients for making *Ubtan* are milk, besan, haldi, lemon juice and petals of rose flower which are easily available at home.The boil leaves of neem, tulsi and rhizome of haldi used as a antibacterial or antiseptic face-wash. The seeds of khus-khus mixed with curd apply to cure dendruf.

Ethnoorthopaedics

It is the ancient traditional way of healing fractures, injury and other disorders of musculoskeletal system. Traditionally in every village a local man is found who deals with settings of dislocated bones and fracture in libs and hands. Many plants which are traditionally used to cure of bone fractures and pain in bones like nirgundi (Vitex nigundo), hadjod (Cissus quadrangularis), atrilal (Peristrophe paniculata). The leaves of nirgundi after crush applied on part of fractured bone in the form lepa to cure bone fracture. Whereas the leaves of this plant also used with alum or fitkari with lukewarm water to cure bone fracture and to reduce the pain.

Paleoethnobotany

It deals with the study of identification and recovary of plant remains from archeological sites by adopting various methods or it is the study of plant relics found in sedimentary rocks. The plant remains may be leaves or fruits or pollens. These plant remains are droped down in various types of gums derived from plants and may be preserved for many years. Others may be impressions on rocks. An expert that involve in did entification and recovary of the ramains is called paleoethnobotanist. Some impotant example of plant fossils which are discovered by Prof. Birbal Sahani are *Williamsonia sewardiana*, *Bucklandia indica*, *Taeniopteris spathulata*, *Pentoxylon sahanii*, *Carnoconites laxum* etc.

Ethnolinguistics

Ethnolinguistics is a study of various types of languages of diffrent ethnic groups that show how these linked to different cultures of socities and their conversations. Language is a main vehicle that transport our old traditions or cultures form generation to generation. Ethnolinguistic study requires datas related to their life style like type of food habit, type of their houses, type of clothes, occupation either a worker or collect non timber forest products (NTFP) from forest and what are their rituals and believes etc. To understand the language or pass out the message of communication it is necessary to study the various types of cultures and languages of various communites.

Ethnopediatrics

It is the study of methods of care taking of health of infant and its mother. Susruta Samhita is the monumental written document on ethnopediatrics anywhere in the world that contains various aspects on childcare like cleaness of newly born infant, its disease, feeding etc. Cow or goat milk or honey is given to newly born infant before feeding of mother's milk in Kol tribe of Chitrakoot. A local house made preparation by using hing, ajwain, chhuara and jaifal is given intermittently to the infant during the six months to protect the infant from common cough and cold. This local preparation is prepared by rubbing the jaifal on stone and keeping the other gradients between jaifal and stone and this is freshly prepared every time when it has to be given. The infant completely depend on mothrs milk for six months. Regarding sleeping of infant, most of the mothers prefer to sleep their infant by right side. The disease jaundice is common in newlyborn infants therefore it is necessary to spent ten minutes under sun in morning daily or alternatively. It triggeres the formation of vitamin-D and altimately makes the bones stronger in infant body. After six moth a programme *Pasni* is organised when infant is given cereal or pulse soup to feed first time. In modern era, this branch is known as paediatrics, which is a Greek word that means science of childcare. In modern medicine or now days, the care provided to the child up to 10-12 years of age.

Ethnobryology

It is the study of bryophytes traditionally used in various cultures or by various ethnic groups. Many bryophytes are used in decoration, packing, bedding and in traditional medicine. The bryophytes like *Sphagnum* spp., *Marchantia* spp., *Weissia* spp., *Funaria* spp., *Plagiochilla* spp., are used to cure eye disease, liver problems, treatment of hair maldness, skin problems, healing of wounds etc.

Ethnodietetics

It is the qualitative and quantitative study of food items that are consumed by the people of vaious cultural groups. The consumption pattern of food in human society is highly influenced by socioeconomics and their culture. This diet varies among the various culture groups and geographical region for example various culture Brahamin community of central India prefer to consume vegetarian food whereas different types of non-vegetarian food are the main component of the traditional food during the festivals and various occasion in in northern India (Himalaya region) in the same community. The quantity and rules for eating of food also vary in various communities like most of the communities inhibiting inside the forest take two times meal early in morning and before sunset at evening.

Ethnohorticulture

It is a traditional practice among various ethnic groups to grow indigenous horticultural crops like fruit trees, vegetables and other ornamental and medicinal plants in their vicinity. In this traditional method all, the plants are grown without using any chemical fertilizer that is now known as organic farming. Therefore, this is very old traditional method in India in which horticultural plants like papaya, mango, banana, guava etc. commonly around the every house in most of the cultural groups. For primary treatment of any disease some medicinal plants like tulsi, sadabahar, maruba, giloi, datura, aswagandha etc are mostly planted near by their houses. Many flowering plants like champa, bela, gudhal, gulab, genda etc may also be observed around their houses and along the village roads or gullies in villages as a old traditional horticulture practice of various societies.

Ethnoveterinary

Domestic animals are the integral part of most of the tribal communities since ages because the livelihood of tribals depend upon the milk, meat and direct sell of these animals in markets and earn money for their various need. Ethnoveterinary is the traditional knowledge or practice to treat various types of disease of common domestic animal. These traditional medicines are chaeaper and easily available in rural or tribal areas therefore, used as low cost alternatives to the costly allopathic medicines. Many plants as a whole or seeds like fenugreek, clover, garlic use to treat the common disease in goat and cattle. Many traditional practices of animal health treatment prevail in India some of very common example such as feeding the dry garins of bajra (*Pennisetum typhoides*) or fruits of gular (*Ficus glomerata*) to the cattles for removal of placenta. Another practice is to give the heat tereatment after the thin layer *Lep* of asafoetida on the navel of animal by using glowing coal after burning of dung cack also helpful in removal of the placenta.

Ethnoopthalmology

It is the study of practices and beliefs of tribals used for caring the eyes or vision. There are many plants which are used against various eye-disease for example the juice of *Boerhavia diffusa*, *Hordeum vulgare*, *Mirabilis jalapa Oxalis corniculata* used in eye inflammation and juice of *Cynodon dactylon*, extract of flower petals of *Rosa* spp. used to treat infected eyes etc.

3

Ethnic People and their Contribution

Indigenous culture

Indigenous means originating where it is found. If knowledge originates from, and is bound to local experiences, taking its local world not perhaps as the only one, but as the most relevant of all they know may be called indigenous. Indigenous knowledge is human life-experience in a distinct natural and cultural amalgamation within a unique local and timely setting. It is an authentic appropriation of being, meaning that this very process happens exclusively in a given locality at a given time. In the field of medicine, Chinese culture (indigenous) developed Acupuncture technique. An acupunturist places needles into the skin to balance out the flow of chi which is believed to be the energy that flows through the human body. In theory, Chi runs along 14 pathways called meridains and when a patient is injured, stressed, or ill, their chi is supposedly being blocked. By placing needles in the Meridians the patients chi is thought to be put back into balance. Like Chinese culture, indigenous culture of different places/countries occurs and have their own belief and methods.

Indigenous cultures speak in various ways, for example we could speak from the perspective of our so-called 'civilizations' but how do we know that this perception is not contrary to the truth. We could speak on the basis of our traditional religious concepts but again these concepts may in reality be the opposite of what we believe. If we speak about the indigenous from the point of view of an anthropologist we arrive at a cold and empty language, a study that only pays attention to external and superficial matters, in the same way an anatomist analyses our inner organs or our bones but they are still a long way from knowing the reality. We believe that it is better to speak about indigenous cultures on the basis of their own reality, in relation to how they were as much as how they are now because in the course of time from yesterday to today many things have happened that are worth examining.

Folklores

Hindi means of folklore is Lok-Sahitya or *Pauranic katha*. The term folklore was coined by William Thomas in 1845. Folklore is the combination of two words folk and lore. Folk means community of people and lore means knowledge and traditions of particular group frequently passed along by words of mouth. Folklores like folk songs, tales, riddles etc. references to certain properties of plants. These uses though not tested in modern laboratories proved correct because of the long years of trial and error method to which they were subjected. Some examples of plant related folklores are given below :

Aconite (*Aconitum napellus*) : Its leaves and roots are extremely toxic and were used by the ancients as a poison.

Apple tree (*Malus domestica*) : The apple tree was an important orchard fruit of ancient Greece. It was associated with love and marriage.

Sage (*Salvia officinalis*) : In London, it is a symbol of immortality. It was often planted on graves in days gone by as it was said to live forever, often thriving on neglect where other plants would fail and die, and it's true that it is one of the easiest plants to grow as long as it is not in wet soil.

Rose (*Rosa domascena*): There are many stories about rose. In Greek mythology about its begning one day flowers Goddess found a weak numph Chloris wish to restore her health and asked to help. The God gifted the joy, charm and beauty. But Chloris wanted to do more for sick and weak nymphsoshepray to other god for giving nectar by Aphrodites. In last Chloris called her beautiful bloom rose. It was believed that all roses were white in begning but it become colorful. According to this Greek mythholology god Jupiter sawVenus bathing, she was so embarrassed that she blushed and all the white roses surrounding her bath turned red in sympathy.

Sandalwood Tree (*Santalum album*): According to Indian mythology, sandal wood tree is depicted as being twined with serpents. Sandalwood remains cool and aromatic even when the poisonous serpent coils around it. This also has another meaning that the basic nature of an individual can not change because of outer effects.

Tulsi (*Ocimum sanctum*): According to Indian mythology, Tulsi plant is most loved by Lord Vishnu and Vrinda Devi, the Goddess ruling Tulsi is known as the personification of bhakti or devotion to the supreme being. Tulsi is considered to be associated with purity and a highly revered and used for all religious purposes among the Hindus. It is considered very auspicious to have a Tulsi plant in the front courtyard of many Hindu households.

Baho-Varuna (*Crataeva religiosa*): The leaves and bark of the tree is pungent, bitter and astringent in nature, used in ailment of various kinds of urinary disorders. It is also used in various tradition uses like restless syndrome, weightloss, urination, immunity, antidote in snakebite etc.

Folktales

Folktales are known as *Lok Kathaen* in hindi. Folktales are old stories originating in particular culture and typically passed on generation to generation by words of mouth rather than stories in written form. Folktales are the soul's nourishment. They keep the flame of possibility alive-Rafe Martin

Folktales related to plants have been discussed here. In our culture folktales have got special place. Every tale involves a tree in one way or another. Trees appear in dreams. They sing and talk. They offer consolation and convey special powers. In many of the tales, a tree serves as teacher or guardian of the truth. Characters, who sit under a tree or climb up into a tree are suddenly inspired to set out on a journey or receive a decisive insight.

1. Why Death is Like the Banana Tree (Madagascar Tale)

God wanted the first man and woman to be able to choose the kind of death they would have. One day he asked, "Would you prefer to die like the moon, or like the banana tree?" The couple did not know what it meant to die like the moon or the banana tree, so God explained, "Each month the moon dies and fades away, but it revives bit by bit to live again. When the banana tree dies, it does not come back, but it leaves behind green shoots so that its offspring can carry on in its place. You may have offspring to take your place, or you may revive each month like the moon. You choose."

The couple considered the options for some time. If they chose to be childless, they would always be restored to life, like the moon. It would be lonely, however, and they would have no one to help them with their work, no one to teach, to love, or to strive for. They told God they preferred to be fruitful like the banana tree. God granted their wish. They had many fine children and a happy life and then they died. Since then there has been much love and new life on this earth, replenishing generation after generation. But since the first couple chose, each individual's life is brief, and in the end the body weathers like a banana tree.

2. The Magic Pear Tree (A Tale from China)

A long time ago in ancient China a farmer went to market. He had luscious pears to sell and was determined to ask a very high price. Once he had found a good place in the market, he cried out: "Pears, beautiful pears...!"

He called attention to his goods, an old ragged-looking monk approached him. He humbly asked to be given one of the pears. The farmer said: "Why should I give a pear to you? You're as lazy as anything and haven't done an honest day's work in your life." As the monk did not walk away but repeated his request, the farmer became more and more angry. He called him the nastiest thing sunder the sun. "Good sir, " said the monk, "I cannot count the number of pears in your wheel barrow. You have hundreds of them. I have only asked for one pear. Why has this made you so angry?"

By then a large crowd of people had assembled around the farmer and the poor monk. "Give him a little pear," some one suggested, in the hope that this might solve the problem. "Do as the old man asks, for heaven't ask it is only a pear," another one remarked, but the farmer wouldn'thear of it. "No is no is no," he said. Finally an elderly man bought one of the pears and handed it reverently to the old monk. The monk bowed, thanked the elderly man and said: "You know that I am a holy man. When I became a monk I gave up everything. I have no home, no clothes which I may call my own, no food other than what is given to me. How can you refuse to give me a single pear when I ask for it? I shall not be this selfish, I invite every one of you to eat one of the pears that I have grown. It shall be an honour if you accept my invitation."

The people were started. Why had the monk asked for a pear if he had so many pears with him? He did not seem to carry anything. What did the old man mean?

The monk at his pear with great concentration until there was just one small peace left. He quickly dug a hole in the ground, planted the peace and gently covered it with earth. Then he asked for a cup of water. One of the people in the crowd handed him the water. The monk poured it on the soil. Hardly any time had passed when the by standers saw some green leaves sprouting from the earth. These leaves grew very quickly. The people were astounded. In front of their eyes stood a small pear tree with branches and more branches and leaves, more and more leaves. Where the old monk had planted the little peace only minutes ago, there was now a small pear tree. It continued to grow faster and faster. They could see it grow.

Silence fell in the marketplace as the tree burst into flower and the flowers slowly turnd into large, sweet-smelling pears. The monk's face was a glow with pleasure. He picked the pears one by one, and handed them to eachperson who had witnessed the peartree' smiraculous growth. He handed them out and handed them out until every one had been refreshed by a delicious pear. Then the monk took his axe and before the people even realized what was happening, the pear tree was cut down. The monk simply picked the tree up, put it over his shoulder and went on his way.

The farmer had watched the scene in amasement. He had not been able to believe his eyes when the pear tree grew out of the ground so near to his very own wheel barrow which was full of pears. He looked at the barrow. It was empty. Not a single pear was left in it. One of the handles of the barrow was missing too. Then the farmer knew what had happened. The old monk had used his pears to create the wonderful pear tree. Of course the monk was no where to be seen. The pear tree which the monk had picked up with such great ease was found a little further down the road. It was the missing handle from the wheel barrow. The farmer was in a towering rage, whilst the crowd laughed.

3. The singing fir (Switzerland tale)

Once there was a master wood carver who lived in the village of Reckingen in the canton of Valais. Many of the churches in this region were beautified by this special handiwork, which was the carving of sacred figures of the saints, and church pews decorated with wooden foliage. One evening when the church bell of Recking rang for the hour of prayer the woodcarver heard strange but beautiful singing. He went to his window to listen and noticed it seemed to be coming from the forest of Hohbach which covered the steep hills that rose above the village. The singing stopped as soon as the bell stopped ringing. "My imagination is running away with me," the woodcarver muttered to himself as he put away his tools and hurried to church. But the next night he heard the singing again. And the night after that he heard it once more. In fact, the strange music began every time the church bell rang for evening prayers and ended when the bell stopped ringing every night for the next several nights. The woodcarver was understandably relieved to discover that many other villagers heard it as well. One evening during prayer time, he climbed the hill to the forest of Hohbach determined to find the source of the mysterious music. He wandered about among the huge trees for a long time. At last the singing led him to a giant fir hundreds of years old. To his astonishment the sound seemed to come out from of its trunk! When it was again quiet he ran to the village to report his discovery. After that, each evening several of his neighbors climbed the hill to the miraculous tree and stood in reverent amazement beside it.

The woodcarver visited the tree often and ran his hands across its bark. The giant fir was constantly in his thoughts, even appearing in his dreams. He became obsessed with the idea of making a carving out of it, to be the most magnificent work he had ever done. The parish finally agreed to cut down the tree and let him work on it. But the woodcutters harbored a secret misgiving as they felled the magnificent fir, and their hearts grew heavy as they dragged it down to the valley with horses. The master carver cut himself a huge block from the heart of the tree. He told the parish he was going to make a statue of the Virgin Mary and set to work that very night. He toiled day in and day out and in his zeal almost forgot to eat. People flocked from miles away to watch the Holy mother slowly emerge from the wood. The woodcarver was truly talented and his work a veritable masterpiece. All said that no artist had ever made Mary look so beautiful and alive.

When the statue was finished, it was presented to the church of Reckingen. The priest took the statue and placed it on that alter as the people watched in silent awe. Then suddenly the wooden figure of Mary opened its mouth, and once more the familiar dulcet tones of the miraculous music were heard.

4. Why plants have human characteristics (Iroquoian Myth)

Before the earth was created there was a land above the sky. Certain beings, men and women with human characteristics but not entirely human, lived among the sky people. These beings grew so numerous that the land above the sky became crowded. The beings began to quarrel among themselves and with the sky people. The sky people went to the Great One and asked: "Can you not do something to bring back peace to this land above the sky?"

The Great one poked a hole in the sky and blew his breath through the hole, so strongly that a cloud of mists formed in the space below. He then asked the sun to shine through the hole. When the sun's rays fell on the mists, they turned to water and formed a great sea. Then the great one called the Moon and asked her to shine through the hole. As she shone down, a thick scum formed on the sea. Gradually, the scum drew together into a solid mass and made the earth, with the sea all around it.

Great one now had a place to send the beings, but when he saw how bare the earth was, he decided he must first change some of the beings into plants and animals, and send them to all the corners of the earth. So Great one changed most of the beings into plants and animals, and then with a great breath, scattered them over the earth. And that is why every living thing on earth has some human characteristic, because each kept one thing from the time when the beings lived in the land above the sky. In animals it is easier to see these characteristics, but if one looks carefully, they are to befound in plants as well. Some plants have leaves shaped like the human hand, or like an eye or ear. Some have hair that looks like human hair. Others have flowers shaped like faces or feet. All of them, whether in an open or a secret place, have one thing that shows they are also descended from the beings in sky, just like the humans.

5. A Girl who was loved by a tree spirit (Nagaland-India)

A beautiful girl lived in a village of Nagaland in India. She belongs to a very rich family. Many young man was invited at night to court her. When she saw a very handsome young men decided to merry with him. He came to her every night and went away before dawn. The girl search him in a day time in near by villages and fields but she could not find. In last she told about this to her parents. On the advise of her parents she tied a thread around the waist of her lover when he came at night. He went on his usual time. When the girl went out in next morning she was surprised, when she found the thread tied around a tree that was standing at walking distance of her house at the bank of a river. Now whole family began to suspect that the handsome young man was not a human being at all but the sprit of that tree. To clarify this doubt the girl put an indigo dyed shawl at next night, on the young man's shoulders same as she had tied the thread. The young man went away in next morning as usual. In the morning they found that shawl was hanging from a forked branch of that tree. It stood near the river where the girl often went for bathing. One day the father of the girl followed this young man and saw this man went straight to the river, he stood at the bank of river and quickly transformed into a ordinary tree. The father of the girl decided to fell down that tree and he called the villagers to help him in felling that tree. The father advised to his daughter to remain inside the house to avoid any mishappening. The people try to cut that tree but the tree would not fall. Than hey start to chopp, a small chip of wood flew towards the girl's house. At that movement the girl was watching the cutting tree through small hole in the wall. The flying chip of wood entered the hole and struck in eye of girl with very high speed and damaged her brain, the girl died instantly on the spot. The girl fell on the floor on the same time the tree was also fell on the ground. Now father was very happy and came back to home with the thought that tree spirit would no longer chase and haunt his daughter. But, he found his daughter dead. The two lovers, girl and tree sprit had died together.

6. The wishing tree (A tale from Rajasthan-India)

Once upon a time a traveller was going through a forest. After passing few miles he felt thirsty, hungry and tired. He wish to restunder a tree. Suddenly he saw a tree infront of him. He was surprised as he was in desert, just few seconds ago there was no even any bush. He was very happy that he has place to rest. He sat down under the tree, thought how I wish I had water to drink. Suddenly he found a pitcher full of cold water kept in front of him under the tree. He surprized again and quinch his thrust and sat down again under the tree. After few minutes the traveller thought how I wish eatsome deliceousood. Suddenly a variety of food appeared before him. He ate as much as he could. After meal, he sat down again under that tree and he though how I am sleeping on soft mats and he found himself on soft mattress. Due to travelling he was tired so he was asleep soon. After a long nap when the traveller woke up, began to think I am alone in this forest could herebe a lion in this forest. Suddenly a lion appeared. The traveller was filled with fear and asked to lion A....A...Are you going to kill and eat up me. The lion replied yes. The lion jumped upon and ate up the traveller.

Protection of wildlife by tribals

The tribal life of India has a long history. The British adopted two policies for them. One was for the removal of certain evil customs, which were prevalent in the society, like sati, child marriage, etc., and the other was the policy of non-interference in the recognition of their customary laws about marriage, adoption, succession, etc. There are many laws for the Scheduled Tribes and at the same time they have their own customary or traditional code of conduct too. The legislative provisions are of two types, one for their welfare and the other that puts restrictions on their conduct.

There are many laws to protect the environment. Some of them are; Wild Life (Protection) Act, 1972, Indian Forest Act, 1927, etc. These acts protect the wild life by declaring some areas as protected areas, reserved forests, protected forests, etc. These acts also affect the tribal laws, which are made for the welfare of the tribesmen. The regulations of many protected areas in the world strictly prohibit hunting, gathering of plant resources or other productive activities with in park or reserve borders. This policy applies to all local people; even those who have traditionally had access to the area and possibly practiced sustainable ways of harvesting natural products from the wild lands. They are forced to continue their activities in buffer areas around the park, which may lack some resources and where they may face competition from enterprises which may be ecologically destructive, such as certain types of logging and agriculture.

The Forest Rights Act, a law recognizing the rights of forest communities, has come into force, with the Union tribal affairs ministry issuing a formal notification to this effect on January 1, 2008. The Scheduled Tribes and OtherTraditional Forest-Dwellers (Recognition of Forest Rights) Act, 2006, was notified after more than a year of political interventions, bureaucratic twists and hectic lobbying by activists representing tribal and wildlife interest groups. The main beneficiaries of the Act are scheduled tribes (STs) and other traditional forest-dwellers living in and depending on forests for their livelihood for three generations or 75 years prior to December 13, 2005. Essentially, the Act aims to provide a framework to record the rights of forest-dwellers, allowing them to cultivate forest land to the extent under occupation, subject to a ceiling of four hectares, the right to collect, use and dispose of minor forest produce, and rights inside forests that are traditional and customary, like grazing and maintaining homesteads. The act renders the Forest Conservation Act, 1980, the Wildlife Protection Act, 1972, and the Indian Forest Act, virtually in valid." The Act also makes it mandatory for right holders to ensure sustainable use, conserve bio-diversity and maintain ecological balance, there by strengthening the conservation regime of forests across the country.

Plant domestication by the tribals

Domestication refers to the process whereby a population of animals or plants, through a process of selection, becomes accustomed to human provision and control. Plants domesticated primarily for aesthetic enjoyment in and around the home are usually called house plants or ornamentals, while those domesticated for large scale food production are generally called crops. Tribes of different parts of world played an important role in domestication of plants as well as animals. These people came to know how to use the plants and what part should be used and what should not. The earliest human attempts at plant domestication observed in Asia. There is early evidence for conscious cultivation and trait selection of plants by pre-Neolithic groups in Syria: grains of rye with domestic traits have been recovered from Epi-Paleolithic time.

Hemp, one of the most ancient plant domesticated in India, is an example of a multipurpose plant: oil is obtained from its seeds, fibres from its stalks, and the narcotic from its flowers and leaves. Some plants were domesticated especially for the production of narcotics; such plant is tobacco, which was probably first used by American Indian Tribes for the preparation of a narcotic drink and only later for smoking. The opium poppy is another example of a plant domesticated solely for a narcotic. In different parts of the world very different species were domesticated.

Taboos

Hindi meaning of taboo is *Nishedh / Varjit*. The taboo label is a product of the human mind. Taboo plants, in general, refer to plants that a certain ethnic group will avoid growing, mostly for in literal sense 'prohibited' or 'proscribed'. It is a prohibition which, if violated, leads to an automatic super natural penalty. There are a number of taboos or 'social ban' related to social customs in various societies. Violation of these social customsis not acceptable in the most of society. Therefore, the taboo is often regarded as 'not allowed set of behaviors' by a society. For example, In Bhil community, killing of squirrel or house cat is prohibited and it is believed that if some one kills the squirrel and cat, they will go to hell. Sarus crane killing is a big taboo in all the three clans (Bhils, Meena, and Khatodi), and they believe that the handicapped child will be born in the killer family. Fishing is also taboo in breeding season by all clans. Certain stones, trees, jungles and rivers can not be touched as these are thought to have supernatural powers and touching these would invite calamities or destruction to the individual or to the society as a whole in many tribal communities.

In Singapore, people avoid growing the *Plumeria*, croton and cordyline inside their home gardens as they have been regarded as graveyard plants. Growing graveyard plants in one's home garden is believed to attract bad luck. Bananas and various night-blooming fragrant plants, especially those with white flowers, are also avoided as they are believed to be linked to spirits.

Taboos which are spread in Monpas tribe in Arunachal Pradesh

The husband of pregnant women should not exchange anything with the outsiders. If they do this there is a chances of exchange of their child in the womb.

The women during pregnancy is not allowed to sleep near the stream or forest, beliving the spirit will take away the child from her womb.

The Monpas are not allowed to eat fresh meat. They have to preserve meat for three days after that allowed to eat. Breaking this taboo may suffer from different types of disease.

A women during menstruation period is not allowed to go and work in Jhum field and to enter in the forest. It is believed that during this period of pollution the women are easily possesed by malignant spirit which bring disease and ailment to the women and her child.

Totems

Hindi meaning of totem is *Kulchinha*. Totem is an object representing animals or plants that serve as an emblem of a group of people such as a family, clan or tribe and serves as a link to their ancestry. A totem is anything in nature that connects it for guidance and wisdom. Fin, feather, and fur often connected to our brother and sister creatures; without failing to see the stones, insects, and plants beneath our feet. Tribal have a unique connection to their totem and how they work together. There are some examples of plant totems.

In India, many tribal groups have totemic trees of their clan, group or tribe. For example in Maharastra, Kol tree (ber) is the totemic tree of Koliya tribe. Peepal tree is the totemic tree of Pimpley tribe. In Karnataka, kadam tree is the totemic tree of Kadambas and kadamb tree is still worshipped as totem by the Gadaas tribe. In M.P. Umar tree is the totemic tree of Umaria clan and semal tree (*Bombex ceiba*) is the totemic tree of Samria clan. Another one, salai (*Boswelia serrata*) is the totemic tree of Salaiya clan. Other than these there are many tribal groups, whose name undoubtedly derieved from different villages has been adopted from the names of various totemic trees like Barodia from bar (*Ficus bengalensis*), Semalwara from semal (*Bombex ceiba*), Khajuri from khajur (*Phoenix sylvestris*), Ambapara from Amba (*Mangifera indica*), Amlikhera from Imli (*Tamarindus indica*) Saagwadeeya from Saagwan (*Tectona grandis*), Semliya from semal (*Bombex ceiba*) and Palaswani from palas (*Butea monosperma*) in Banswara district of Rajasthan. In Chittorgarh district of same state, the village Dhawarakudi from Dhawra (*Anogeissus pendula*), Anwalheda from Anwla (*Emblica officinalis*), Kanthariya from kanthar (*Flacortia indica*), Barkhera and Barodiya from bar (*Ficus benghalensis*), Nimbaheda from Neem (*Azadirachta indica*) etc.

Peony: Peonies are a wonderful flower found in many gardens in May. Grown best in shade and moist composted soil they grow from a small stalk into a huge delight of color and smell. Peony seeds were often dried and carved to wear around the neck for protection to travelers and prevent evil curses. Peonies planted around a home grant protection and health to all who lived within. Peonies also speak of learning how we may start small in the world but grow into a huge success. Peony energy is strong yet nurturing and loving, reminding us to nurture ourselves and our desires so that they may bloom forth into our daily lives.

Red Clover: Red Clover and all the Clover family have been with humans since we first herded beasts to graze in fields. Often considered a farmers friend it was fed to the cattle as food, boiled and stained for a sweet tea or face wash, and use to aid cough and inflammations. Clovers are everywhere humans are found. Clover seeds can live 50 or more years in the soil until the time is right. Clover also brings oxygen and nutrients into the soil allowing for other plants to seed in. Clover promotes life, generosity, abundance, and sweetness to all that walk barefoot through her cool leaves. The spirit of Clover is playful and calls to us to remember the hours spent looking for that special four-leaf clover. When we relax and look for the small details is when we open ourselves to the Fairy Realms. Clover patches often grow in a circle creating a doorway between the Fairy/Spirit world and ours. So next time you pass a clover circle take a moment to look. Search for that lucky four-leaf wonder and watch for the little white lights of Fairies saying hello.

Carnations: Carnations are a common favorite at the flower stand. They come in every color in the rainbow and can be grown with ease in the garden. We receive carnations at happy events or give carnations to friends that are sick. Carnations' spicy smell helps us to focus and visualize our dreams and desires. White carnations for memory, focus and creation. Blue and green carnations for growth, new beginnings, and easing of emotions. So go to your local flower stand and choose a carnation today. Let it's smell and color bring love and health to you and your dreams.

Weeping Willow: Lazy summer days by our favorite pond or lake is embodied in the spirit of the weeping willow. Willows are a common tree found through out the world. Their love of water was a sign to deep water below. Their flexible branches are used for dream catchers to form the frames. Their calming effects are also found in the aspirin we take for headaches. So next time you need to relax or meditate try sitting next to a weeping willow. The soft sound of her leaves will put you into a trance and relieve the peace you're looking for.

Daffodils: Daffodils are one of the first flowers to appear in the spring. Their bright colors give a promise to the warmth of summer to come and the blooms of other flowers. Although they come in a mixture of colors; yellow still remains the favorite of many. Their bright color of yellow is one of the most healing vibrations. In combination with their bright green stalks daffodils bring healing and prosperity to any home or office. The color yellow brightens moods, and allows us to see the world in a different light. Their strong thick leaves protect them from frost and their flowers can turn toward the direction of light. Allow the spirit of daffodils to turn your heart toward the warmth of cheer and healing.

Magico-religious therapies

There is a general observation that in most cases the people do not worship but react to crisis situation caused by supernatural agencies. The people perform rituals in order to appease such supernatural beings so that they withdraw their wraths and malign activities. These magico religious therapies or practices are grouped here in to four zones namely, northest, central, east and southzone of India. The zone wise detailed description is as follows:

North zone

Natural therapy

The use of plants for treatment in India dates back to prehistoric times. This indigenous knowledge about medicinal plants and therapies was composed verbally and passed orally from generation to generation. Much later, some of this information was systematized in treatise forms like *Atharveda*, *Yajurveda*, *Charak Samhita*, *Sushrut Samhita*, etc. These systematized systems of knowledge about medicinal plants and therapies are included under Ayurveda. Despite the development of rural health services, village people still use herbal native medicines to a large extent for treatment of common ailments like cough, cold and fever, headache and body-ache, constipation, dysentery, burns, cuts and scalds, boils and ulcers, skin diseases and respiratory troubles, etc. The herbal medicines are prescribed by the household ladies, elderpersons, Pujari, Ojhas etc., and by traditional herbalists.

Jagar

In magico-religious therapies, they practice Jagar, Thau-dham, Bhabhuti, Tantra-mantra, etc (Gupta *et al.*, 2014) to placate the local gods and supernatural powers. And in natural therapies, like in Ayurveda, they use herbal products. According to the mode of application, the natural therapies have three categories:

1. Herbal products used in systematized system of medicine like Ayurveda, Siddha.

2. Herbal products used in ethno-medicine or indigenous medicine like HMS based on oral tradition.
3. Herbal products used in modern medicine, based on active chemical principles of the herbal products.

The Himalayan people believe that diseases are caused by unhappiness of local gods. Thus they treat diseases through some magico-therapies. Mostly the magico-religious physicians are called *Pujari*, who are the mystic-priest of a village. The Himalayan people use some native medicine but if a person does not recover from an affliction, his relatives approach the mystic-priest (*Pujari*). The *Pujari* tells them whether the patient is under the spell of an evil spirit or has incurred the anger of the local god, or whether he is suffering from some sort of illness. In the latter case, the patient is taken to a village herbal physician for treatment. If the *Pujari* decides that the patient is under the spell of an evil spirit, here commends some other mystic-priest, with the help of hymns, drives the spirit away. Before leaving the patient, the spirit may ask for some rice and pulses or for the sacrifice of a cock, pig, goat, or some colored cloth. The rice and pulse are left on road crossings. It is believed that the spirit will leave the patient after getting the articles demanded. Finally the *Pujari* puts some ash marks on patient's, forehead which is locally called as *Bhabhuti*. If the *Pujari* says the patient is under the anger of the local god, he recommends a magico-religious ceremony known as *Jagar* to placate the god.

Jagar is always held at night. A large room in the patient's house is cleaned and provided with articles of worship such as fruit, cereals, milk, curd, ghee, sweet, etc. It is decorated with various flowers and branches of some special trees. The *Jagaria*, who is the hymn chanter and conductor of the ceremony, the *Dangaria* (the dancer who acts as a medium for the appearance of the god), and the relatives and friends are seated in ceremonial room. A crude drum (*nagara*) and a metallic plate (*thali*) are played, the *Jagaria* chants hymns, and the *Dangaria* begins to dance. When the dance and the music reach their climax, the patient's house hold god speaks through the medium of the *Dangaria*. The relatives ask the god-in-medium the cause of his anger. He tells the cause, which may be, among other reasons, that the patient did not worship him properly or that he did not give him a sufficient share from his earnings. As a penalty the god may demand a simple *Khichari* (a mixture of uncooked rice, pulses, chillies, and salt); a sacrifice of a goat, pig, cock, or coconut; a continuous *Jagar* for 20 days or so; or simple worship at home or in a particular temple. Everyone has to fulfill the demands of the god because failure to do so many result in serious consequences not only to the patient but also to his family. One may witness such a magico-religious ceremony in any village of the Himalayas, even among the educated classes. About 50% of the patients are cured by it. These medical systems are psycho-somatic in nature and need to be properly investigated.

Bagai dadi regad regad adi

This magico-religeous practice is performed by various communities living inside forest in Garhwal region of Uttarakhand. Here, people were used the seeds of castor for protecting their domestic animals from the tiger by this practice. In this practice, they kept the seeds of castor mixing with duoble number of small stones at the evening of Deepawali festival and used to call 'bagai dadi regar regar regad'. There is a believe this practice de sharpen the teeth of tiger.

Northeast zone

1. Monpas tribe of Arunachal Pradesh

The Monpa tribe of Arunachal Pradesh is considered as one of the major tribal community in the northeast. Monpas are the inhabitants of the high altitude Tawang district and the mountain passes of Bomdila in west Kameng district. Traditionally Monpas tribe fears from God or supernatural powers and they believ that every novel activity took them a step towards spiritual gain and upliftment (Baishya, 2009). The following magico-religeous practices are pravellent to appease supernatural powers and dities related to different disease for treating the disease and evil spirit.

Prayer

In Monpas tribe, an elaborate type of prayer is practiced. They use various types of prayers which depend upon the seriousness of crisis. The following disease are cured through prayer offer to lord Buddha according to the direction of priest.

Disease: Headache, vomiting and stomach problem

The priest performs dhutpa prayer in front of Dang-dang diety where, they light earthen lamp and after over the prayer asked the patient to smell it.

Disease: Madeness and senselessness

The priest performs phan prayer ritual in front of doot diety. Where, red colour small flags are fixed towards the jungle during the prayer.

Disease: High fever, malaria and blood vomiting

Phan prayer ritual is same but it is performed in front of Tsan diety. Where, red colour small flags are fixed opposite to the patient's house after the over of ritual.

Dokkgyur

For the recovery of Luspogom (body pain) a ritual is performed which is known as dokkgyur which is performed in the guidance of the priest. In this ritual, a prayer is offered to Lord Buddha in monastery and incant on the entire body of the patient. This is repeated till the patient recovered from the illness.

Dripthroy

Dripthroy is performed for treatment of paralysis in monastery. Ingredients like milk of white cow, red cow, milk of goat, and water from stream collected in the early morning to performing this ritual. The priest set fire to dammar and requested the patient to inhale the smoke of the dammar. The priest made an amulet without disclosing the name of material used for making amulet and gave it to the diseased person.

Mey-ngah

Mey-ngah ritual is performed by Monpas for the burnt cases in monastry. In this ritual the priest, pray infront of Lord Buddha and offers ghee with the prayer. Ghee after offering is given to the patient to apply on the burnt area. The burnt person put this ghee until the comlete recovery.

Sang-sol

This ritual is performed for the treatment of Jaundice by the Monpas priest. The priest given a water to the patient for drinking as a medicine to appease Lord Buddha in monastery. The patient of jaundice maintained this ritual until the full recovery.

Sheen Dokthap

The ritual sheen dokthap is performed for the treatment of fracture of bones. Initially the priest covered the effected part with paper and paste with pinjar made out of skin of animals in water. After giving coverup to the effected part priest, chanting on the affected part of the body with prayer in front of Lord Buddha.

Thori

The magico religious therapy thori is performed when the patient is suffering from headache for a long time. In this therapy the priest offers a prayer to Lord Buddha and chanting on the forehead. This practice is repeated after the interval of two nights until the full recovery of patient.

Zadrip

This therapy is used when the patient suffered with epilepsy. In this priest offers a prayer in monastery three times in a day at the gap of three days. Ingredients like milk of white cow, red cow, milk of goat, and water of a stream are used to cure epilepsy.

Ngan

This therapy is performed when the patient is suffered diseases like skin disease, abnormal delivary etc.

Phurdo

This is performed to remove the evil spirit from the ailing person. Before performing this therapy, the priest fix five colourful flags in front of house of the ailing person. The priest of monapas made some idols of rice flour and ghee and offer it flower. Priest light the ghee lamp infront of this rice flour idol and appease the bed spirit. This therapy is performed upto three days at the house of patient.

2. Zeme-nagatribe of Assam

It is a group of Naga people from northeastern part (Assam) of India. Most of the traditional Zemi rituals are performed as protective and anticipatory measures to prevent economic failure and to overcome crisis (Roy, 1998) from disease and sickness which are as follows:

Paulia

This ritual is performed every year in the month of February for the general well being and good health of the community of Zeme Naga.

Meurokappe

To protect crops from evil influence of a spirit called Mekong and it is performed in March.

Habaikeu

The whole community of Zeme Nagas for economic prosperity performs this in the month of September as a religious festival.

Nteugumeurokappe

This ritual is performed in September at community level for protection of agriculture crops from Mekong. Mekong is appeased for good and safe harvest of the agriculture crops.

Hipumra

Hipumra ritual is performed by community as well as individuals in January every year for good health of the people.When it is performed at community level, called hipumra and when it is performed at individual level, called kilungra. At community level the prayer is offered in temple by observing a particular day that is known as tingna. During the tingna the main entrance of village is closed to stop the entry of evil spirit and outsiders inside the village. People of the same village are also not allowed to go outside the village till the complete of the ritual. Celebration of hipumra is compulsory to all the community members while celebration kilungra is not compulsory to all.

Hemui Kakelumbe

This ritual is performed in month of march to get rid from evil power or bad thoughts and fear. Utterance of evils by such persons can induce ill effects on the victim's body and mind. There is a perception about this ritual is like leaves of trees shade off in march same all the evils or ill effects go away from the life of the people after performing this ritual in march.

Paiki

Paiki ritual is performed by paipeu equivalent to a priest. All young one of the family are the members of the village paiki and knows a paikiana members. Paiki comprises two words pai means cleanliness and ki means hut. It means it denots to a clean hut ofr house. During the performing of paiki prayer any unclean visitors like women during menstruating cycle or animal like pig dog etc. are not allowed inside the house.

East Zone

1. Santals of Orissa

The ethnological characters of the santals distinguish them from all other races in India as nomadic and civilized. Santals called themselves 'bir syndic' or strongman and 'horhopon or son of the man. They are remarkably well made with strong limbs, thick lips and dark skin. Their hairs mostly long and coarse, tied in a knot on the top of the head. They do not have any written literature though their traditional legends are current among them. Santals have socio-religious and economic importance of trees of sal, Mohua, Haldu/Karam, etc. for various kinds of NTFP in their day to day life. Like other tribals Santals are polytheists. They believe in number of dieties, ghosts and the spirits resides in the hills, forests and streams. His benevolent supreme diety is Sing Bonga who is identified with the sun and worshipped with great reverence as the creature of universe.

Bhujni Batiyani

Bhujni-Batiyani ritual is practiced in the Santal community of Sonatangri in Orissa (Chakrabarty *et al.*, 2012) for well being of the villagers. This practice is performed to protect them. Ritual Bhujni Batiyani is practiced by the Santals of Sonatangri for well being of the villagers. They believe that by observing this ritual they will be able to protect and keep themselves away from the effect of all epidemic diseases. Bhujni, also known as '*Jugni*' among the Santals of West Bengal, is a malevolent female deity existing in the Santal pantheon where, "Batiyani" means leaving apart. Thus the ritual "Bhujui Batiyani" stands as leaving apart Bhujni from the village. This ritual generally takes place in the month of April which according to Santal almanac is known as Chait Bonga. But the day and date of the observation of the ritual is not fixed. For this purpose, on the penultimate night of the day of the ritual, the unmarried boys of the village assemble at the residence of the majhi (traditional village headman of the Santal). And on that night respective Santal household of the village collect their used and rejected brooms, baskets etc, and keep them in front of their respective houses. In the early morning of the next day, the unmarried Santal boys of the village collect all these dumped materials from every household and tie those materials in carrying pole. There after, the boys carry those materials along with a chicken and enter in to the near by forest. Throughout the day, they collect different medicinal plants, roots, fruits etc. from the forest. For the entire day time they spend in the forest and it is prohibitory for them to enter even within the boundary of their village. In the forest, they allow the chicken to be free with the belief that it will carry all the disease from the village and transport them to the forest. After spending a whole day outside the village boundary they return in the evening. However, before entering in to the village boundary they sacrifice three chickens and dedicate them respectively to the Marang Buru, JaherAayo and MorekoTuruiko (God and Goddesses of the Santals). After the episode of sacrifices is over, they go straight to the residence of majhi. In the premises of the majhi they grind all the collected medicinal plants, flowers, fruits; roots etc together and make a dry mixture out of those grounded materials. Thereafter, they distribute the portion of the mixture to all the Santal household of the village.

South India

1. Irular tribe of Tamil Nadu

Rangam

The Irular tribe mostly dominated in Tamil Nadu, Kerala, Karnataka and Andhra Pradesh. Rangam is a magico-religeous practice of Irular tribe. The people of Irular tribe consult the diseases and health condition of the village. Rangam performed by Shaman a sacred specialist of the Irular community. At the time of Rangam, Shaman is possessed by the diety of Kanniatta that is the mother Goddess, who announces through the medium of his wife, who are the spirit or demons and what will satisfy them? After knowing the bad spirit the Shaman suggests the remidies to get-rid from the bad spirit.

4

Methods and Tools in Ethnobotany

Before carry out any ethnobotanical work, first define the goal of the assignment or study before selecting the approach, which best suits the interests, budget and schedule. Most of the methods used in ethnobotanical studies are time taking and costly because several field trips are required in several seasons for covering different ritual occasions and seasonal farming activities which are continue for a number of years for completion of an assignment. In this way, an assignment may be very long or short as per selection approach of an ethnobotanist. Long-term assignment or projects may be continue for several years and they require sufficient financial input. When sufficient financial add and time does not available to accomplished long term assignments, in that case a Rapid Ethnobotanical Appraisal (REA) can be done.

Rapid Ethnobotanical Appraisal (REA) for short term projects

This technique is appropriate when we have to complete an assignment in a short time without requiring expensive tools because participants seek a sketch of local conditions rather than an in depth study. In REA a small group of native community is selected for interviewing in a semi-structered way covering wide range of topics.This wide range of topics gives a complete detail of a particular community. Although REA have some limitations like it is not allow deep working relationship between ethnobotanist and community, incomplete documentation of the cultural and biological aspects of local knowledge. Short time visits do not permit local people to learn rigorous ethnobotanical methods etc. Documention of the traditional knowledge of any area may be done in a short time and investing least money by adopting REA. Rapid ethnobotanical appraisal can be prepared with the help of the following points:

- Secondary source of information (maps, floras, faunas, vegetation analysis, census statistics)
- Fomation of team including local people (including botanist, anthropologist and linguistic expert)
- Consent of local community (Corporation and permission of local authorities before starting fieldwork)

- Systamatically arrange working plan (appraisal should be arrange systematically by including maps of site, name of infomants, identification of plant species and conclusions of the appraisal, together with any drawings or charts).

Long-term Projects

After completion of REA of any targated area, the research may be continued for several years as a long-term project to document the ethnobotanical knowledge of the targated area.When we decide to carry out a long-term project we should describe the landuse, local people and conservation status of the region where the research project will be execute.The geographical location of the research site should be stated, giving the country, state, province, municipality etc. The history of the geological formation of the land should be described as well as the topography, soil types and geographical landmarks (mountains, caves etc) and human population size, language and local dialect speaks, non-timber forest products extracted, and floral and faunal account of the area. Demarcate the major ecological zones, indicating seasonal fluctuations in temperature and precipitation. The documentation of ethnobotanical knowledge during long trem projects may be completed by considering the following points:

- Identification of plants used in their tradition system and information collected on the distribution, use and management of the corresponding botanical species.
- Herbarium samples of all plant species represented in the study must be collected, identified and deposited in a herbarium.
- Local plant names and other terms must be accurately recorded using a widely accepted local writing system.
- The local perception including concepts of vegetation communities, soil types, geographical landmarks, climate zones and seasons should be recorded.
- The economic value of biological resources and availability of plant and animal products sold in rural market, assess the time people spend in harvesting the resources should be estimated.

Collection of ethnobotanical data

Ethnobotany is a field-oriented course involving direct relationship of primitive societies and surrounding plants. The ethnobotanical data may be collected by adopting various methods, which are as follows:

Field Work

Ethnobotanical fieldwork is different from routine plant collections for floristic studies which are done by taxonomist. A taxonomist give emphasis on plants, ethnobotanist in addition record record the various uses of plants by local people. So first step is to search out a local medicine man or tribals, develop familiarity with them and other knowledgeable informants in literature, culture, dialects, vegetation etc.Tribals by nature are very conservative and shy, and do not like to share the knowledge with outsiders, even with some incentives. Medicine man tend to secretive in presence of their own tribe, whereas they are fully cooperative when asked in private. Knowledgeable informants are taken to the field, along with collections of plants; the use of plants as given by the tribal informants is recorded.

Ethnobotanical fieldwork needs researchers to use integrated approach with government, scientific community and local people. Researchers must obtain official permits before conducting any research work, especially when conducting research in indigenous communities. Field workers should also obtain permission, compensate the informants with a some amount of money and intellectual property rights regarding the ecological knowledge of the indigenous people must be taken into consideration before initiating research activities and then coduct his research work with honestly. Target of local informants is also a common mistake is assuming that one is dealing with culturally homogeneous groups; it is often a considerable degree of intracultural variation.

Plant identification and collection

Plant collection is a most basic work in ethnobotany. Collections are worthless because they use as voucher specimens, which are permanent records of the plants recorded in a certain place or community. A representative plant sample should has flowers, fruit or both to make the identification easier. Collected plant speciemen to be keep inside airtight polybag to stop scenacence and subsequently shading of leaves. After coming back in camp, the sample specimens should be pressed inside blotting paper using herbarium press for drying. The herbarium label should include the name of the university/organization, name of collector, family, genus and species of the specimen, specialist and date of determination, locality, vegetation and habitat, latitude and longitude, altitude, plant description and collecting date.

Field notebook must be prepare while making plant collections and not at the end of a day or trip in order to avoid the loss of data. A standardized numbering system should always be used to label all collections with the field notes. Local plant names contain interesting information about the use and perception of plants by a particular community. Fieldworkers should be familiar with the local language and work with the cooperation of a particular language of a community. All the local names of plant and their products should be recored with taking samples of that plant for correct scientific naming of the lantspecies. Now a day there is an era of digitization so it may be documented by using a photo camera also. Pictures of a plant in its natural state can be very useful for its correct identification.

For collection of ethnobotanical data, field surveys are conducted. These are two types of field surveys (i) Questionaaire survy: In which questionaaire is used for gathering information from a broad range of literate informants usually in urban and semi-urban areas. In this type of survey the respondant fill the quationaaire alone, which means that there is little or no contact with researcher (ii) Interview: It is a set of questions usually typed or hand written on a paper. These forms are field by researcher after face to face interaction with lierate or non-literate informants.

Mostly four types of questions are used in ethnobotanical surveys, open-ended, dichotomous, fill in the blanks and multiple-choice. Dichotomously questions are mostly answered in yes or no, true or false, agree or disagree. Fill in the blanks type of questions include with six openers why, who, what, how, when and where. For example, why this plant used against a particular disease, what is the local name of the plant etc. In multiple choice questions the informant given a limited number of possible answers like, this plant is used as fodder, fuel, medicine or ornamental. Whereas, open ended questions might be use for at any stage of research with very high information content covering the broad subject area and low number of querie and ease of statistical analysis.

Interviewing

The main way of gathering ethnobotanical information is to talk with people, to watch what they do and to participate in their activities. The various field interviewtechniques are given by Alexiades (1996) and Martin (1995) which are as follows:

Field Interview

This method is very time consuming because the interviewer or ethnobotanist walks in the field with the informants. In field, he listens to the informant and writes a note on the various uses of the plants.

Participant observation

This technique is based on manand plant realation viz. homegarden management. In this technique, the participant consists of walking in the field or in the forest with an informant, listen to him and asking him about plants and gathering fruits, other forest products, and writing notes on their uses.

Plant interview

In this method, plants are collected from the field and bringing back to forest finge villages and ask about the uses of that plant from the local people. This method is very useful in reconfirmation of the ethnobotanical use of a plant species.

Artefact interview

This type of inetrview is very good for small and primary ethnobotanical studies because it is simple and will familiarize the local community with the ethnobotanist. In this researcher asks the Informants about the plants, which are employed in the construction of the house and preparing, tools, baskets, etc.

Checklist interview

The ethnobotanist compiles a list of plant names and present it to the informants. This technique is suitable for common plant. The local name of any plant can vary from village to village or a group of a community so this is the limitation of this technique.

Group interview

In this technique, the interviewer conducts interviews with a group of informants. Group discussions can produce a good amount of data and lead to new questions. Some people will be more willing to share their knowledge in a group environment, while others will be hesitat to disclose certain types of knowledge in front of others.

Mapping

According to Townsley (1996), the mapping is a topographic distribution of the plant specimens, homegardens, households and local infrastructure. Maps, GPS records taken during the survey and aerial photographs give the basic topographic information. The latitude, longitude and altitude of the collected plant specimens should also be notedwith the kelp of GPS. It is clear that rough maps of the area should start drawning with starting of discussion with local people in any survey. These maps give a physical plateform for discussions and something concrete for people to refer to when talking about local conditions, changes in conditions and allow them to explain their view of their environment and relevance in it for them. Mapping helps to complete limited information provided by existing maps, which are often incomplete, and to get a better understanding of key local features. Finally, they provide a basic sketch map for use for filling in thematic information, as historical changes in settlement, social structure, socioeconomics and land use pattern of a particular tribe or inhabitants.

Scrutiny of herbarium specimen in herbaria

According to Schultes (1963) the ethnobotanical data are anchored to herbarium sheet makes easier the problem concerning the proper identification and its ethnobotanical information, plant parts used their processing, preparation and doses etc. For the large collections in any herbariumare scrutinized for ethnobotanical data from one place. Excellent work in this field is done by Altschul (1968). He scrutinized thousands of thousands of herbarium specimens in Harvard University Herbarium and recorded notes on 5178 uses of plants. But in India very little work has been done on this regard. Very little work has been reported by Jain and Dam (1979). So this field has immense scope for ethnobotanical research in India.

Scrutiny of ancient literature

Old literature provides very useful data about plants used by people of those days in different parts of the world. Unpublished literature may be historical chronicles, diaries and travelogues by adventures part from the mention of plants used, life and culture of people of those days are highlighted. In India vast heritage of Vedic literature like Ramayana and Mahabharata have numerous plants names and its role in the daily life of people of those days.One of the best examples of the use of ancient literature in ethnobotany is the large compilation of anti-cancer plants cited in the old text and local folk medicines. Atharva Veda, which is another classical work, describes about 2000 plants having medicinal properties. Sushruta Sarnhita (1000 BC.), further records the medicinal qualities of 700 plants. But there are so many constraints in ethnobotanical studies from ancient literature viz., correct identification of the vernacular names with the latin names because local names mentioned in ancient literature vary from place to place and time to time. Ex. Brahmi, Soma, Jatamansi etc.

Archaeological remains

Study of archeological remains gives important information about ethnobotany of past time. These remains include hard coated seeds, charcoal, timber etc. which buried along with made structures long back ago. Excavation of these archaeological sites and plant parts reveal certain information on new sources of medicine, food, fibres etc. In some studies the wild races of food grains cultivated in those early periods have also been discovered. Archeoethnobotanical investigations help in finding out migration of human cultures, origin, dispersal and domestication of agricultural crops. Sitholey (1976) reported about 40 plant sps. from great Stupa of Sanchi and the railing of Bharhut Stupa belonging to the first and second century B.C.

Scrutiny of ancient sculptures on temples, forts and paintings

Sculptures on temples and forts are the important source of ethnobotanical knowledge.Mrs Shakti Gupta worked on temple sculpture and painting and published a series publications on vast use of plants by people of old days.

Analysis of different tribal folklore

Folk songs and folk tales which are passed from generation to generation have some intresting information of plants. These informations though scientifically not tested and may give knowledge after careful analysis.

Credibility analysis

Various tribal people start the use of different natural resources around their habitation areas. After observation, trial and error of long experience they discovered the use of various plant and animal products, which is an important part of indigenous knowledge. The indigenous knowledge that got tested and accepted by numerous workers, recognized as a Ayurveda, Sidha, Unani system of medicine. Several workers suggested numerous methods for evaluating credibility of such folk or traditional knowledge.

Information collected after the interviews of any tribal area are authenticated by consulting other members and only those information which are found the same are recorded. Balick (1996) given a methodology named as 'credibility rating' to find out the authenticity of the information collected on the ethnobotanical use of different plants.

Table 1: Credibility rating for use information collected

No.	Category	Credibility Rating
A	Collecter uses or directly observed use	1
B	Informant uses or directly observed use	2
C	Informant heared or knew from further source	3
D	Use reported from the literature	4
E	Common knowledge	5
F	Credibility of use information unknown	6

For credibility of ethnomedicinal plants which are used in ailment of a disease in more than one region or state or by more than one tribal ethnic group point criterion may be used. Before use of this credibility criterion the plant species associated with ethnomedicinal claim in more than one state or reported more than one ethnic group are arranged alphabetically by scientific name than each plant has given some point of credibility in the scale of 1-5 on the basis of species in one or different localities, region, states and ethnic groups which is also suggested by S.K. Jain (2004) which is given below:

Table 2 : Credibility criterion based on multilocational and multiethnic use

Criterion	Point
More than one report, but from same locality	1
Two reports from different localities or from two different tribes within a state	2
More than two reports from two adjacent state but different ethnic groups	3
Two reports from distant states and ethnic groups	4
More than two reports from different states and ethnic groups	5

Relative Cultural Importance Indices (RCI)

Gausset (2004) told that various RCI indices pool the particular uses quoted by informants as a "use categories". There are endless potential categories and subcategories, whereas the standardized categorization requires thorough compilation, comparison and efficient presentation of the ethnobotanical data. The common subcategories are merged in to broad categories such as construction, food, medicine, firewood. In his book to explain varios RCI Indices Phillips (1996) to describe various RCI indices, given three broad methods including: "uses totaled", "subjective allocation" and "informant consensus".

Use Value (UV)

For calculation of Use Value (UV) for each species methodology of Farreiraet al.(2009) and Phollips et al. (2002) is very useful and moslyuse in calculation of ethnobotanical data. Use Value evaluates the relative importance of each medicinal species based on its relative use among informants. It is calculated as:

UV= ΣU/n

Where, U is the number of times a species is cited and n is the total number of informants for a species.

For example, Interviews with two informants concerning one plant species, each informant was interviewed twice (Events I & II). Each specific use is tallied, although specific uses can be grouped in use categories. For each species, the total uses cited by an informant, is summed and divided by the number of events to arrive at the UVis (use value of the species for a single informant). The average of all the UVis for a species is the UVs (total use value of the species for all informants).

	Informant 1				**Informant 2**				**UV**
	Construction	**Medicine**	**Food**	**Total**	**Construction**	**Medicine**	**Food**	**Total**	
				Species 1 Shisham (*Dalbergia sissoo*)					
Event 1	2	6	6	8	5	0	2	7	6
Event 2	2	0	0	2	3	0	4	7	
UV	2	0	3	5	4	0	3	7	

Cultural Value Index

Cultural Value Inex (CV) is developed by the Reyes-Garcia et al (2006) which is calculated with the help of the following formulae

$$CV_e = Uc_e * Ic_e * \sum IUc_e$$

Where *e*: ethnospecies (voucher specimens did not match up with herbarium botanical species so were referred to as ethnospecies),*Uce*: number of uses reported for an ethnospecies di-vided by the total number of use categories. Ice: Number of participants who listed a species as useful divided by total number of participants. ÓIUce: Num-ber of participants who mentioned each use (catego-ry) for the ethnospecies divided by the total number of participants.

Practical Value Index

$$PV_e = Up_e * Ip_e * DUp_e$$

Where *Upe*: number of uses reported for an ethnospecies divided by the total number of use categories. *Ipe*: number of times an ethnospecies was brought for use divided by the total number of participants. *Dupe*: an assignment of duration of use for each item brought to the household. Randomly selected subjects were asked about the plants they brought home within a 24 hour period.

Fidelity level

Field Level (%) = (Ip/Iu) x 100

Here, *Ip* indicates the number of informants who independently designate the useof a species for the same major ailment and *Iu* the total number of information who mentioned the plant for any major ailment (Friedman et al.,1986).

E.g. Total Number of informants (Sample size)= 40Disease Malaria and plant species used is *Andrographis paniculata*

Number of informants cited the plants: 14 *(Ip)*other than malaria the plant is used for other diseases: 21 *(Iu)*

FL= 14/21x100=66.66 % Contd

Proforma for field work given by S.K. Jain and A.K. Goel as follows

Form A

Place
Serial No.
Recorded by
Name of Informer
Date
Tribe
Sex
Age
Remark

0.1 Name of the Tribe
0.2 General description
0.3 Polulation
  0.3.1 Size, distribution
  0.3.2 Lowest density- Population/total area
  0.3.3 Economic density-population/ cultivable area.
0.4 Location.
  0.4.1. Name of place, village, Distt., State.
  0.4.2. Physical Feature (Hill top, Valley, slope, plain, vicinity of some river, mountain etc.).
  0.4.3. Weather
  0.4.4. Facilities of camping
0.5. Environment.
  0.5.1. Rainfall and Humidity.
  0.5.2. Temperature.
  0.5.3. Wind storms.
  0.5.4. Sunlight.
  0.5.5. Soil.
1.0 Cultural settings.
  1.1. Area under forest.
  1.1.1. Description of forest.
  1.1.1.1. Evergreen.
  1.1.1.2. Mixed.
  1.1.1.3. Deciduous.
  1.1.1.4. Scrub land.
  1.1.1.5. Grassland.
  1.1.1.6. Riverain.
  1.1.2. Major forest produce (Give briefly: details to go in Form B)
  1.1.2.1 What articles collected chieflyTimbersFruitTreesOthers
  1.1.2.2. What price fetched.
  1.1.3. Minor forest produce.
  1.1.3.1. What articles collected.
  1.1.3.2. Price fetched if any.
  1.1.4. Misc. forest produce.
  1.1.5. Any forestry price.

1.2. Area under cultivation.
1.2.1. Agricultural produce.
1.2.1.1. Major crop list.
1.2.1.2. Minor crop list.
1.2.1.3. Agricultural calendar (Dates)
Sowing harvestingother operation
1.2.1.3.1 Jan.
1.2.1.3.2 Feb.
1.2.1.3.3 Mar.
1.2.1.3.4 Apr.
1.2.1.3.5 May
1.2.1.3.6 June
1.2.1.3.7 July
1.2.1.3.8 Aug.
1.2.1.3.9 Sept.
1.2.1.3.10 Oct.
1.2.1.3.11Nov.
1.2.1.3.12 Dec.
1.2.2. Agricultural operations.
1.2.2.1. Main operations with local names.
1.2.2.2. Main tools used.
1.2.2.3. Any rituals with agric. operations.
1.2.3. Agricultural labour.
1.2.4. Transaction in Agr. produce.
1.2.4.1. Within community.
1.2.4.1.1. How marketed.
1.2.4.1.2. Price.
1.2.4.1.3. Barter.
1.2.4.1.4. Credit.
1.2.4.2. Outside the community.
1.2.4.2.1. How marketed.
1.2.4.2.2. Price.
1.2.4.2.3. Credit.
1.2.5. Agricultural property rights.
1.2.6. Shifting cultivation.
1.2.6.1. Sowing and harvesting.
1.2.6.2. Fallow lands.
1.2.6.2.1. Why left.
1.2.6.2.2. What grows on these.
1.2.6.2.3. Any utilization of this vegetation or totally abandoned.
1.2.7. Agricultural improvements.
1.3. Occupation other than agriculture.
1.4. Animal Husbandry.
1.4.1. List of animals.
1.4.2. Feed requirements of animals.
1.4.3. Crop damage.
1.4.4. Manure.

1.5. Cultivable waste.
 1.5.1 Area.
 1.5.2. Why abandoned.
 1.5.3. What possibilities.
1.6. Non-cultivable area.
 1.6.1. Area.
 1.6.2. Why non-cultivable.
1.7. Historical account of changes in occupational habits if known.
 1.7.1. During last 10 years.
 1.7.2. During last 100 years.
 1.7.3. Relation with any conspicuous change in environment.
 1.7.4. Relation with any social or political change.

Proforma for recording information on plants used by tribal people

Form B

Place State, Distt., Village.................. Reporter..................

Tribal or family...................................... Date......................

Informer Name......................................

Tribe/ Caste..........................

Age.................Sex.............

Plants Usage

0. Timbers

0.1. House building.

0.2. Tools.

 0.2.1. Tools in agriculture.

 0.2.2. Tools in hunting.

 0.2.3. Tools in collecting wild foods, etc.

 0.2.4. Tools in weaving.

 0.2.5. Tools in spinning.

 0.2.6. Tools in dyeing.

 0.2.7. Tools in painting

 0.2.8. For woos work.

 0.2.9. For basketry.

 0.2.10. For pottery

 0.2.11. For fishing.

 0.2.12. Tools in forestry

0.5. Furniture etc.

0.6. Boat building.

0.7. Toys and carving.

0.8. Bridges.

1. Food.

 1.1. Normal human diet.

Local name	Bot.Name	Quantity Consumed	How Consumed	When consumed	Source Grownor wild	Cost now procured	How preserved

 1.2. Special diet

 1.3. Any notions of relation of diet with Physiology

 1.3.1. With fertility, sterility or abortion.

 1.3.2. With intelligence or dullness.

 1.3.3. With activeness or sluggishness.

 1.4. Misc. actions about food.

 1.4.1. Offerings to God before meals.

 1.4.2. Bathing or any other action essential before meals.

1.4.3. Posture while eating.

1.4.4. Any particular action after food, i.e. smoking, sleep, rest etc.

1.5. Fodders.

1.5.1. Cultivated.

1.5.2. Wild.

1.6. Any foods offered to deities.

1.6.1. As a usual practice.

1.6.2. On festivals.

1.6.3. Distributed as Prasad of God's.

2. Plants used in medicine:

Local Name	**Bot.Name**	**Partused**	**How medicine prepared**	**Method used**	**Source**

3. Dyes and Tans:

3.1. Dyes in foods.

3.2. Dyes in clothes.

3.3. Dyes in houses.

3.4. Dyes in toys and timber articles.

3.5. Misc. uses.

3.6. Tannins used for shoe leather.

3.7. Misc. tans.

4. Fuel and light.

5. Waxes, detergents.

6. Perfumes, oils.

7. Resins, gums, mucilage.

8. Insecticides.

8.1. In agriculture.

8.2. In domestic hygiene.

8.3. In community hygiene and sanitation.

9. Paper and pulp.

9.1. Any paper manufacture done

9.2. Any toys etc. made of pulp.

9.3. Any other use of paper pulp.

10. Beverages.

10.1. as narcotic and drug

10.2. As diet.

10.3. As medicine.

10.4. How prepared

10.5. Rituals connected with.

10.6. Any transaction of sale or purchase.

11. Miscellenaeous

    11.1. Fertilizers.
12. Fibres used.
    12.1. Dresses for males.
    12.2. Dresses for females.
    12.3. Dresses for children.
    12.4. Colour of dresses.
    12.5. For dress of Gods
    12.6. For drapery.
    12.7. For dress of domesticated animals.
    12.8. For cordage.
    12.9. For mattresses.
13. Plants used for ornaments.
    13.1. Natural cosmetics.
    13.2. Natural colours.
    13.3. Natural beads.
14. Any other unusual utilization or application of plants.

Proforma for data recording on Sacred Groves

Name of the sacred grove ....

Ownership: (Temple Trust/ Revenue/ Forest Department/Panchayat/ Private)...

Legal status (whether protected legally as sanctuary/ other any category of protected area)

Name of God/diety of sacred grove ....

History of sacred grove ....

Area ....

Date of study ....

District ....

Tehsil ....

Block ....

Name of nearby village ....

Name of gram panchayat ....

1. Location
   1.1.Latitude
   1.2.Longitude
   1.3.Altitude
2. Climate
   2.1.Rainfall
   2.2.Temperature
   2.3. Wind
   2.4. Humidity
3. Soil
   3.1.Soil Type
   3.2. Soil organic matter
   3.3. Soil pH
   3.4. Soil EC
   3.5. Soil bulk density
   3.6.Extent of soil erosion
   3.7. Protective measures against soil erosion (if any)
4. Socioeconomic data of nearby village
   4.1. Name of four local people of nearby village
   4.2.Total population of village
   4.3. Male/female ratio of the village
   4.4. Occupation of villagers (Primary, secondary)
5. Rituals
6. Folkflore
   6.1. Main event related to sacred grove (Festival/ annual event/ special event)
7. Ethnobotany
   7.1. Name of the tribe
   7.2. Name of the tribal headman
   7.3. Name of the tribal medicine man
   7.4. Name of the plant family

7.5. Local name of the plant
7.6. Locality of the plant
7.7. Growth stage of the plants for medicinal use (leaf/flowering/fruiting etc.)
7.8. Description of plant
7.9. Availability of plant
7.10. Any special use of the plant by the tribe (tradition/ceremonies/beliefs/food etc)
7.11. Ethnobotanical herbarium no.
7.12. Plant collected and identified by
7.13. Chemical constituent of the plant
7.14. Therapuetic properties of the plant
7.15. General Description of forest type

8. Biodiversity/Natural resource of sacred grove
8.1. Tree species of sacred grove
8.2. Shrub species of sacred grove
8.3. Herbaceous species of sacred grove
8.4. Climber species of sacred grove
8.4. Epiphytes of sacred grove
8.5. Important flora of monsoon
8.6. Important flora of winter season
8.7. Major fauna of sacred grove
8.8. Major weeds of sacred grove
8.9. Regeneration status
8.10. Water source (Well/handpum/river/pond)
8.11. Water level of area
8.12. Dependance of villagers for fuelwood/NTFP

9. Threats to existence of sacred grove
9.1. Forest fire or insect attack
9.2. Extent of enchroachment
9.3. Illegal grazing
9.4. Illegal cutting and felling trees
9.5. Construction activity (roads/railway track etc.)
9.6. Invasion of any particular weed

10. Conservation measures for sacred grove
10.1. Observation of villagers on health status of sacred grove
10.2. Any restoration efforts taken by villagers
10.3. Plantation inside sacred grove
10.4. Social forestry plantation ousidesacred grove
10.5. Fencing around the sacred grove
Photographs of the sacred grove

Any other remark

5

Plants in Mythology

Plants have played an important role in many mythologies and religions of the world and have been known for deep and sacred meanings throughout the ages. It is the symbolic representation of the old cultural traditions in the form of a tree across the world. Christmas tree in the Christianity and Bodhi tree in Buddhism are the example of tree featured in mythology. In most of the folklore, folktales and folk religion trees are often said to be homes of tree sprits. If history is traced, we will found that Druidism and Germanic paganism appear to have involved in religious conservational practices of sacred groves. Druidism is a nature based religion that was taught by Druids about 2000 years ago. The term *druid* itself possibly derives from the religeous word for oak. Therefore, the Druidism reconnects us with nature, our ancesters and ourselves by working with plants or trees.

Trees are honored in Indian culture for their different beneficial characters like medicinal properties, food, fodder, maintaining hydrological cycle, moderate the climate etc. There are some trees that have attained more than a special benevolence in Hindu mythology. In Indian mythology number of plants have been given sacred value and worshipped by different cultures. Some important plants of sacred value have been discussed here.

Aam (*Mangifera indica*)

Common name: Mango tree

Family: Anacardiaceae

The mango tree is one of the sacred tree of the Hindus. The importance of this tree is also mention in the *Puranas and Ramayana*. The fruit of mango is a sign of love and fertility. The leaves of mango tree are used during religious and communal ceremonies in the Hindu religion. A *"Purnakumbha"* is the sign of Mother Earth, the mango leaves symbolizes life, water is the life giver and coconut divine consciousness. Mango leaves are also used to adorn entrances at home to signify good fortune during auspicious occasions. Mango flowers are also used on *Basant Panchami* for the worship of Goddess Saraswasti. The mango tree is also sacred to the Buddhists because it is said that Lord Buddha performed the instantaneous creation of a large mango tree from the seed at Shravasti.

Mangifera indica: tree

Mangifera indica: flowers & fruit

Amla: (*Embelica officinalis*)

Common name: Indian gooseberry

Family: Euphorbiaceae

Dhattrika or Dhatri which means earth or mother particularly the nursing mother, perhaps because of its nourishing juicy fruit is a tree associated with both Shiva and Vishnu. In Brihaddharma Purana there is a mention about the sacredness of the tree both to Parvati and Lakshmi, the wives of Shiva and Vishnu respectively. According to the legend, both the goddesses went to Prabhasa, a sacred place to worship. Goddess Parvati said to Lakshmi that she wished to worship Lord Vishnu through a new material. Goddess Lakshmi replied that she also wants to worship Lord Shiva through a new substance. At that, the two goddesses wept and from their tears were created Dhatri or the Amalakitree, which both propitiated to worship Shiva and Vishnu and since then the leaves of this tree are considered essential in the worship of Shiva and Vishnu. The tree is worshipped on Shive-ratri day. It is surrounded by red and yellow thread, flowers and fruits are offered to the tree just as in any other worship. It is considered as one of the most sacred trees in Gujarat. In the month of Kartik (Oct.-Nov.) women worship the tree with flowers, sandal paste and sindoor, particularly on the 9th day of Kartik, called Akshaya Navami when women worship it for be getting male progeny. They make five round of the tree and tie sacred thread in each round on its stem.Amalaki is a plant, which transformed itself into a beautiful woman to entice Vishnu away from the charms of *Vrinda* (*Ocimum sanctum*). The tree is very sacred to the Hindus and credited with magical properties by the tribesmen. This plant, which is of great medicinal value, is planted on the South side of the temple or home.

Embelica officinalis: tree

Embelica officinalis: fruits

Anjeer (*Ficus carica*)

Common name: Fig tree

Family : Moraceae

Fig tree is associated with Lord Buddha and it is believed that Buddha have attained enlightenment under this tree. The leaves are effective against ringworm and scabies. Fruits have antidiabetic and anticancer property.

Ficus carica: plant in flowering

Ficus carica: fruits

Arandi *(Ricinus communis)*

Common name: Castor bean

Family: Euphorbiaceae

It is said that *Swayambu lingam* appeared under this shrub at Koteswarar temple at Tirukottaiyur in Tamil Nadu. The use of castor seedoil in India has been long back for use in lamps and in local medicine as laxative. In Manipur, the leaves of this shrub are heated and applied to women's breast to increase milk secretion. In Garhwal, castor uses for pacifying and pleasing Shukra Graha.

Ricinus communis: plant in flowering stage

Ricinus communis: fruits

Ashoka (*Saraca asoca*)

Common name: Sita Ashok

Family: Fabaceae, Subfamily: Caesalpinioideae

The tree Ashoka is one of the religious and well-known trees of India with very beautiful perfumed flowers. These flowers and foliage of the tree are used in temple decoration and in many other religious celebrations. Ashok tree, is revered by the Hindus, Buddhists and Jains. Literally meaning 'sorrowless tree'. In India, people believe that drinking the water in which the flowers of Ashoka tree have been dipped is widely considered a protection against grief. It is believed that Lord Buddha was born under the Ashoka tree at Lumbini, in Gaya. But according to another group, Lord Mahavira renounced the world under tree of Ashoka in Vaishali. According to Hindu mythology this tree is dedicated to 'Kama Deva'. It is also mentioned in 'Ramayana' in the form of Ashok Vatika, where Hanumanji first time meets with Sita.

Saraca asoca : tree

Saraca asoca : flower

Banyan *(Ficus benghalensis)*

Common name: Banyan tree

Family: Moraceae

The name banyan given to *F. "benghalensis* come from the Gujarati language where banyan means merchant. A tree where Hindu merchants conducts their business. This word frequently used by Purtagess for Hindu merchants and passed it to the English. English writers began to write this tree as banyan tree. Banyan tree is the symbol of the three lords Vishnu, Brahma, Shiva.In Hindu culture, this tree is also the symbol of fertility. Due to this belief, banyan tree is worshiped by those who are childless and they never cut this tree. The canopy of this tree may be cover several hectares of land. Giant Banyan in the Indian Botanic Garden, Howrah, is considered to be the largest tree in the world. Lord Dakshinamurth, symbolizes Lord Shiva and seen as the the destroyer of ignorance and embodiment of knowledge, is usually depicted beneath a banyan tree.

Ficus benghalensis : tree

Ficus benghalensis : fruits

Bael *(Aegle marmelos)*

Common name: Wood apple

Family: Rutaceae

In India, Bael tree is considered as very sacred because it is related with Lord Shiva. It is believed that Lord Shiva is pleased by offering leaves of bael tree, also known as bilva tree. The Brahmanas worshiped Lord Shiva for a period of one fortnight by offering bael leaves and that way satisfied Lord Shiva. The fruit, flowers and leaves of the tree are all offered to Lord Shiva. Planting these trees around home or temple is make holy and equivalent to worshiping a Linga with bael leaves and fruits. The trifoliate leaf of the bael tree is believed to symbol of the three eyes of Lord Shiva of three "functions creation, preservation and destruction. It is mendatory to offer the leaves in any ritual or while worshipping Lord Shiva in India. The tree of bilva is also holy to the Jains. It is said, the 23rd Tirthankara, Bhagwan Parasnathji attained *'Nirvana'* enlightenment under the tree of bilva. Moreover religious significane, almost all parts of the tree have medicinal properties and the products obtained from this tree are highly nutritive. Moreover, this also acts as environment purifier by absorbing various poisonos gases from environment. This tree is also worshipped by womens because *Sat Savitri* saved the life of her husband under this tree and this worship is known as *Vatsavitri vrit.*

Aegle marmelos : tree

Aegle marmelos : fruits

Bans *(Bambusa vulgaris)*

Common name : Bamboo

Family: Poaceae

The bamboo is a symbol of friendship. The common names of bamboo are Lord Krishna-Venugopal, Bansilal, Murali and Muralidhar reflect the association with Bansuri or Venu, in the form of steady companion. *Bansuri* is actually a flute made of bamboo. This is the reason; bamboo is honored in India because it is associated with Lord Krishna. In astrology, bamboo is a representative of Punarvasu Nakshatra. In Hindu religion, bamboo is a symbol of family progeny and burning of bamboos means destruction of future generations that mounts to *Pitradosha.* About the use of bamboo, it is said that bamboo is used from 'Cradle to coffin' means start of life and coffin means end of life indicating bamboo use from birth to death. The Ayurveda medicine is one of the ingredients of Banshlochan, which is the most popular formulation for pregnant women and the stem of bamboo used to making funeral bed.

Bambusa vulgaris: plant *Bambusa vulgaris*: flower *Bambusa vulgaris* : seed

Brahm Kamal (*Saussurea obvallata*)

Common name : King of Himalayan flowers, sacred Saussurea

Family : Asteraceae

Brahmkamal is associated with Brahma the God of creation. The white flower of brahamkamal bloomes after sunset and takes about one and hour to full bloom. The home where the flowers bloom believed to be very lucky and sign of prosperity. This flower is offered in shirnes of Kedarnath and Badrinath in Uttarakhand. In Hindu mythology people said that God showered Brhamkamal during the revival of Lakshmana younger brother of Lord Rama using *sanjivanibooti*. Whereas other people believed that Brhamkamal is emerged from the naval of Lord Vishnu. This flower is also related to Hindu God Ganesha. The people of upper Himalayan region of India and Tibet use this plant as a good appetiser and excellent liver tonic. Other than these it is use in treatment of cough, cold, fever, urinary infections etc.

Saussurea obvallata : plants with flower bud *Saussurea obvallata* : flower

Kela (*Musa paradisiaca*)

Common name: Banana

Family: Musaceae

Despite the fact, that banana is not a tree but it is considered a tree outside the premises of botany of forestry because of its shape and size. It is a holy tree and all parts of the tree are used for various purposes. Like, the pseudo stem of banana is used to erect welcoming gates. The leaves are used to make the ceremonial pergolas during Lord Satynarayan Katha. The leaves are used to serve *'prashad'* after different worships or poojas. It is said that offering of the leaves of banana pleases Lord Ganesa. The fruit of banana is offered to Lord Vishnu and Laksmi. Actually, the eleventh day of the *UjaarPaksh* (bright half) of *Pausa* is considered to be very fortunate to offer banana to Lord Vishnu and Goddess Laxmi and sixth day of the *Ujaar Paksh* (bright fortnight) of Kartika is considered fortunate to offer banana to the Sun God. In some parts of India, banana tree is worshipped while fasting on *Vrahaspativar*. On this day, fasting people wear yellow clothes and offers yellow fruit like banana. According to tradition, during Vaisakha, Magha or Kartika suklachaturdasi, a banana tree is planted and nurtured till it start bearing fruits. It is a believe that worshiping the banana tree with flowers, fruit will help in bringing happiness and peace to the family.

Musa paradisiaca : plant

Musa paradisiaca : fruits

Bhang *(Cannabis sativa)*

Common name: Hemp

Family: *Cannabinaceae*

In Hindu mythology, the Bhang is a holy plant. Many beliefs are associated with the bhang plant. Acording to believe, a guardian lives on the bhang leaves. To see the plant or leaves of bhang in a dream, considered very lucky as it brings richness and success in life. If someone meet with you with carrying bhang is a sign of victory. Bhang is a well-known drink made of the leaves and flowers of the bhang plant and considered to be a *"prashad of Bholenath"*. It is must for every disciple to have bhang on *Shive-chaudash*. There is a beliief that nothing good can come to the man who trample underfoot the holy bhang plant. Since ancient times, Most of the Yogies take long draughts of bhang centering their thoughts without any disturbance because bhang has intoxicating propertity. Bhang has also many medicinal properties. It is also believed that no god or man is as good as the saintly drinker of bhang. It is also believed that to restrict the use of such a holy and gracious herb as the hemp or bhang would cause widespread agony and frustration.

Cannabis sativa : plants

Cannabis sativa : flowers

Nariyal *(Cocos nucifera)*

Common name: Coconut tree

Family: Arecaceae

In Sanskrit the coconut palm is also known as *"Kalpavriksha"*, that means '*wish-fulfilling tree*'. The fruit of nariyal is said to symbolize Lord Shiva and the three black marks on the coconut shell, denotes his eyes.This tree has a special place in Hindu religion and great care is given to narture the tree. In the southern India, every household plant coconut trees and is used for all religious purposes. Infact, it is used in the '*sthapana*' of any holy event or pooja. A pitcher filled with water, mango leaves and coconut, that is known as '*Purnakumbha' or* '*Kunkarsh'* is a symbol of Goddess Laksmi and the coconut represents godly consciousness. Breaking a coconut in the beginning of any event or ceremony is regarded to be very auspicious.

Cocos nucifera : tree

C. nucifera : fruits

Kamal (*Nelumbo nucifera*)

Common name: Lotus

Family: Nelumbonaceae

It is a symbol of beauty, purity, faithfullness and divinity and a highly revered flower in Hinduism and Buddhism. Many deities of Hindu's are pictured sitting upon a lotus or holding a lotus flower in their hand. Goddess Laxmi is also always portrayed as sitting on a top of fully opened flower of lotus. Likewise, God Brahma is depicted as emerging from lotus white colored kamal flower is the only plant to fruit and flower concurrently. The flower of kamal is a symbol of *Mata Laxmi* who is the goddess of wealth and fortune. She is the incarnations of the wife of the God Vishnu in Hindu mythology. It is believed, that goddess Laxmi was born radiant and fully-grown from the churning of the sea. The breaking of the surface every morning suggestive to desire and this lead to being associated with spiritual enlightenment.

Nelumbo nucifera : plants

Nelumbo nucifera : flower

Neem (*Azadirachta indica*)

Common name: Margosa tree

Family: Meliaceae

On the first day of Chaitra, after Amavasya, it is necessary to worship the neem and eat its leaves, mixed with pepper and sugar, as a safeguard from fever in Hiduism. The neem tree is a highly revered tree among the Hindus because it is a sign of 'Goddess Durga'. Neem tree also have various medicinal properties. The neem tree is intensively worshiped in various parts of the India. Ladies in Tamil Nadu dance in public places swishing the branches as an act of exorcism and to purify the environment. The Goddess Yellamma assumes the appearance of a young neem tree. In Bengal, neem is the abode of the great pox-mother who can cause or cure chickenpox disease. The neem leaves are rubbed on the body of patient as a traditional practice of treatment of pox, while making prayers to *Sitala Mata*. It is also said that the twigs and leaves of neem are also used traditionally in neutralization of snake venom.

Azadirachta indica : tree

Azadirachta indica : flower & fruit

Peepal (*Ficus religiosa*)

Common name: Sacred fig

Family: Moraceae

As mentationed in Hindu mythology, there were two demons namely Peepla and Ashvattha. They would lure the innocent people to touch the tree and they could kill them. However, later on they were killed by Shanidev on Saturday; this is the cause people worship the peepal tree on Saturday. As mentioned in the Brahma Purana and the Padma Purana, when the Gods were defeated by demons, Lord Vishnu hid himself behind a peepal tree. Therefore, it believed that the peepal tree is a symbol of Vishnu. According to another belief, the tree represents the Trimurti-the roots being Brahma, the trunk Vishnu and the leaves Shiva. According to another belief, Goddess Lakshmi also inhabited the tree, especially on Saturday and hence worshipped for prosperity. In seek of wealth; people light the earthen lamp under the people tree on every Saturday. This tree is also called'Bodhi tree' because Lord Buddha is believed to have attained enlightenment under the peepal tree. Therefore, it is sacred tree in Buddhism also.

Ficus religiosa : tree

Ficus religiosa : fruits

Rakta Chandana (*Pterocarpus santalinus*)

Common name: Red Sandalwood tree

Family: Santalaceae

It is said that Lord Ganesha was created by Goddess Parvati out of sandalwood paste that was used for her bath and breathed. It is used in fragrance industry, fine wood working and aromatherapy etc.The heart wood of chandan is commonly used for incense and religious ceremonies. It is said that all other trees are considered ordinary trees and are like ignorant men in front of a raktachandana. It has an extraordinary fragrance. Sandal wood paste is used in all religious rituals. The paste is smeared on the foreheads of followers of Vishnu and Shiva. There is a believed that the sandalwood paste is meant to cool and protect the 'Agnachakra' between the eyebrows. It is believed lucky; the death pyre is made using sandal wood.

Pterocarpus santalinus : tree

Pterocarpus santalinus : flower & fruit

Tulsi (*Ocimum sanctum*)

Common name: Holy basil

Family: Lamiaceae

In Hindu mythology, tulsi is known by various names like Vrinda, Vishnavi and Haripriya. Tulsi is symbol of purity and used in many religious purposes among the Hindus. There are two species of tulsi plant viz. Rama tulsi, which can be easily identify with help of green leaves and whitish-green stem. Second one is Shyama tulsi with purple green leaves and purple stem. Both Rama and Shyama are the principal avatar of Lord Vishnu. It is very fortunate to have a tulsi plant in the front courtyard of any house because of its sacredness and purifying quality of environment. Person who waters and cares the tulsi plant daily believed to gain *Moksha*. It is also mentioned in *Padam Puran*, a person whose funeral pyre is done with tulsi twig gains *Moksha* (*Vakunt*). All Yogies and Mystics wear a necklace of tulsi beads around their necks in India.Leaves of tulso also an impotant component of *Panchamrit*.

Ocimum sanctum : plant

Ocimum sanctum : flowers

Champa (*Plumeria alba*)

Commonname: Temple tree, Pagoda tree

Family : Apocynaceae

The generic name of *Plumeria* came from French botanis Charles Plumier. In India, there are two common species *P. alba* with white flowers and *P. rubra* with red flower. Champa tree has extra ordinary capability to produce flowers and leaves even after it has been uprooted hence it is a symbol of immortality in many religions. This tree can be seen planted near the temples, Gurudwara, Masjid and graveyard of Muslim community. Hindus and Buddhists offer the flowers at the temples and Muslims to the tomb. Buddhists use the wood of *plumeria* for carving the image of Lord Buddha. The flowers of champa are the colour of a silver moon; whose fragrance overwhelmed at night and grew near the graveyards and temples both. As said in Viatnamese myth, ghosts live on this tree with white and fragrant flowers and white colour is associated with death and funerals. In Hindu mythology, there is a saying about why honey bees do not visit on flowers of champa '*Rooptajey to Radhikey, bhanwar krishana ka daas, esmaryaada ke liye bhanwara aye paas*' meaning here beauty of champa is compared with Goddess Radha and honey bees are servants of God Krishna.

Plumeria alba : tree

Plumeria alba : flowers and fruits

Barna (*Crataeva nurvala*)

Common name: Sacred Barna

Family : Capparidaceae

It is believed that burna tree is associated with Lord Shiva and the leaves are offered in worship on Mahashivaratri. Astronomically this tree sacred to Rahugrah (Neptune). That is one of the nine planets or the navagrahas. It is used for curing urinary infections, rheumatic pain and to reduce the swelling.

Crataeva nurvala : tree in flowering stage

Crataeva nurvala : flowers

Doob grass (*Cynodon dactylon*)

Common name: Bermuda grass

Family : Poaceae

Doob grass is known as durva grass in Sanskrit. There are two words *Duhu* and *Avam* that makes the word durvaor Duhuavam means far away and brought closer. It indicate sdurva brings distant spritual Lord Ganesha closer to us. In Hindu belief durva grass is sacred to Lord Ganesha. According to Hindu mythology, The few drops of drink *Amrit*a felld on the grass during snatching Amrit vessel by *Deva* and *Danava.* Gaining Amrita drops the durva grass became immortal. Durva grass has strong regeneration power therefore it is a symbol of rebirth, fertility and prosperity. Some people says durva grass is the hair of Sita entered in the earth. Rama tried to hold on to her but catched few hairs of Sita and these hairs turned into durva grass. Other says that Sita placed this sacred grass between herself and Rawana during her confinement period in Sri Lanka. This grass makes the line that prevent Rawana from crossing the line of propriety.

Cynadon dactylon : plant in flowering stage

Cynadon dactylon : Seeding stage

Daheya *(Streblus asper)*

Common name : Tooth brush tree

Family : Moraceae

According to Hindu mythology daheya tree is related to Lord Dharukavaneswarar at Tamil Nadu. It is believed that the Lord emerged in the form of a Swayambu lingam under daheya tree. The reason behind giving the name Tooth brushis its twigs are used as a toothbrush for strenghtening gums and teeth because of its antibacterial properties. Its various parts are used in treatment of Toothache, filariasis, dysentry and cancer etc. Latex posses healing property of cracks in the hands and heels and eases swelling of the lungs when taken internally. The leaves of the daheya tree are used as sand paper in traditional wood working in many countries including Vietnam.

Streblus asper : plant in flowering stage

Streblus asper: fruits

Devdar (*Cedrus deodara*)

English name: Himalayan Cedar

Family: Pinaceae

The name Devdar derived from Deo means God and Daru means wood, it is a wood of God. In Hindu mythology deodar tree is worshipped as devine tree. Essential oil of the plant used in treatment of the ailments of dandruf, bronchitis, cough, common cold and osteoarthitis etc. This tree is also well known to Egyptians for its use in preserving mummies.

Cedrus deodara : tree

Cedrus deodara : cones

Nimbu (*Citrus limon*)

English name: Lemon

Family: Rutaceae

In Hindu mythology, the lemon fruit is associated with Lakshmi twin sister of Lakshmi. Alakshmi is the goddess of powerty and sorrow whereas Lakshmi is the goddess of wealth and happiness. That's why nimbu and mirchi are hang at the door of house to stay out the Lakshmi outside the house. It is also said that Lord Shiva appeared under a lemon tree. The lemon fruit is also used after cutting and reciting few mantras for get rid of the bad spirits.The fruit of lemon is a rich source of vitamin C, which enhance the immunity of human body.

Citrus limon : plant in flowering stage

Citrus limon : fruits

Kadamb (*Mitragyna parvifolia*)

Commonname: Kaim, true kadamb

Family: Rubiaceae

Kadam tree is related to Lord Krishna, who climb on a kadamb tree on the bank of river Yamuna and play a wooden flute to confuse and surprise his mother Yashodha. Kadam tree is also associated with *Chirghat* at Vrindavan where Lord Krishna stole the clothes of Gopis during her bath and brought it on kadamb tree that is still there at *Chir Ghat*. Many of Lord Krishna's *Leelas* are said to have taken place under a Kadam tree. Whereas in southern part of India, the Kadam tree is associated with Goddess Parvati. Who is believed to dwell in a Kadam forest (Kadamb khandi-vasini). The tree is also referred as Parvati's tree. The flower is also sacred to Lord Kartikeya. Jainism followers also worship kadamb tree. BhagwanVasupujay, the 12th Jain Tirthankara is associated with the tree.

Mitragyna parvifolia : tree in fruiting stage

Mitragyna parvifolia : flowers

Karanj (*Pongamia pinnata*)

English name: Pongamtree

Family : Fabaceae sub.f. Paplionaceae

Karanj tree grows in coastal region along the streams and rivers. It is associated with Goddess Varahi, the consort of Varaha, incarnation of Vishnu. Various parts of Karanj tree are used in treatments of the disease related to the eye, skin and the vagina. The oil is used to cure tumours, wounds, ulcers, itching and other aliments and natural pesticide all over the world. Due to pinnatified leaves it is called pinnata which are also used as cattle fodder. It is also helpful in reducing environmental pollution by collecting dust on their broad leaves.

Pongamia pinnata : tree in fruiting stage

Pongamia pinnata : fruits

Khejri (*Prosopis cineraria*)

Common mame : Sami, Chhokar, Jhand

Family : Fabaceae, subfamily : Mimosaceae

Khejri tree is also known as queen of desert. It is mythologically important tree of Bishnoi community of Rajasthan and Haryana. When Bishnois begin construction of well they organize a worship under this tree. On the occassion of Shri Krishana janmashtmi they worship the branch of Khejri tree in regard of their diety Gogaji. Some people said that the the Pandavas hid their weapons under khejri tree during fourteen years exile. This tree is also associated with Chipko movement in 1730 AD. This was a environment protection movement where 363 people sacrifice their lives for saving Khejri tree in khejrli village of Jodhpur Rajasthan.

Prosopis cineraria : tree

Prosopis cineraria : flowering stage

Imli (*Tamarindus indica*)

English name : Tamarind

Family : Fabaceae, Sub family: Caesalpinoideae

The word Tamarind come from Persian word Tamar-I-Hind (date palm of India) and the word Amla taken from Sankrit word Amlika means sour taste. Common folkflore of tamarind tree, it is a home of spirits that do not let under the tree survive. Therefore, travelers are advice not to sleep under the tree. There is a belief in singer community, if anyone eat some leaves of very old and giant tree standing near the Tomb of famous singer Tansen, make their voice very sweet as Tansen. Another belief is that when the king Dashrath of Ayodhya exiled his son Rama to forest for fourteen years, Rama went to forest with his wife Sita and younger brother Laxmana. After reaching in forest, they built a small hut under the shade of Imli tree. The tree had large leaves and these sheltered the hut from sun and rain. However, Rama was not happy and he conveyed to Laxmana our father sent us in this forest to see how we can coupe up with adverse conditions in a forest, but you built a hut under very good shelter. By listening this Lakshmana took, his bow and shoot arrows in to the canopy of Imli tree. The large leaves of Imli tree splitted in to smaller one so that could not

Tamarindus indica : Tree

Tamarindus indica : flowers

protect them so that these leaves are remained splitted till date.The tree is best for its fruits, which are used in a variety of delicacies. Various parts of tree are used to cure diabetes and intestinal ailments.

Tejbala (*Excoecaria agallocha*)

English name : Blinding tree

Family : Euphorbiaceae

Tejbala tree is also related to Lord Shiva or Natarajan. There is a mythology about Natrajan temple in Tamil Nadu. There was a Shiv eshrine existed here and it was known as Thillai. The sap of tree used in fish poisoning and arrow poisoning by the local inhabitants. The different parts of the tree are used in traditional medicine to cure sores, tumour and bacterial diseases.

Excoecaria agallocha : plant in flowering stage

Excoecaria agallocha : fruits

Chandan (*Santalum album*)

Common name : Sandal wood tree

Family : Santalaceae

Goddess Lakshmi resides in sandal wood tree. The heartwood of this tree is offered in worship to Lord Shiva as mentioned in *Vamana Purana*. It is second most expensive wood in the world. The aroma of sandalwood oil is esteemed by Hindus, Muslims and Buddhists. It is an important ingradient of havana in Hindu religion. In Buddhist religion the sandle wood is burnt in prayers and meditation. Chandan oil calm the mind and give strenghth, courage and happiness. It also have medicinal properties and used as aphrodisiac, expectorant, sedative and diuratic etc.

Santalum album : tree

Santalum album : flowering

Lasora (*Cordia myxa*)

Common name : Indian Cherry

Family : Boraginaceae

Lasora tree is associated to Lord Shiva and the forests of lasora are discussed in Ramayan as *Sleshmaataka Vana*. The leaves, bark and fruits of the tree are used in treatments of variety of disease. The fruits of the lasora tree are used in dry cough, chronic bronchytis, influenza etc. The bark is used in treatment of fever and skin disease. The leaves are used in wound healing, ulcers, headache and treatments of sleeping sickness.

Cordia myxa : tree

Cordia myxa : fruits

Katahal (*Artocarpus heterophyllus*)

English name : Indian Jackfruit tree

Family : Moraceae

The katahal tree is related to four arms Lord Ganesha and the fruit is associated to Lord Vishnu. The Leaves are used in treatment of ulcer, diarrhoea, stomachache etc. The fruit pulp have cooling effect and aphrodiasic in nature. The ripe fruit is also used as "Brain tonic. The seeds are diuretic and wood is used as antidiabetic and spcially pith said to be abortificient. The root of katahal tree are used in treatment of asthma, fever and skin related diseases.

Artocarpus heterophyllus : tree

Artocarpus heterophyllus : fruits

Madaar/ Safed Aak (*Calotropis gigantea*)

English name : Crown flower

Family : Apocynaceae (Earlier Asclepiadaceae)

The shrub Madaar is associated with Lord Ganesha. Hindus worship Lord Ganesha on Ganesh Chaturthi by offering the leaves and flowers.

The plant is closely linked with Suryadeva. According to the some historical literature, the safed aak plant emerged from spilled milk during a sacrificial ritual to the Sun that was named Arka. Hindus kept the leaves of madaar on their head at the time of holy bath on Rathasapthami. The tincture of leaves of madaaris used in fever as an antipyretic, latex and root bark used as anti-malarial.

Calotropis gigantea : plant in flowering stage

Calotropis gigantea : flower

Mogra/Bela (*Jasminum sambac*)

Common name : Jasmine

Family : Oleaceae

The jasmine plant is associated with Lord Vishnu. Whereas the flowers of jasmine are the symbol of love and romance. In Goa, jasmine festival is conducted in Mardol Mahalsa Temple every year in August-september month and jasmine flowers are offered to Devi Mahalsa during worship. More then three lakhs rupees of flowers are used for decoration of the Mahalsa temple. This plant also have various medicinal properties like sedative, antiseptic, aphrodiasic, expectorant etc.

Jasminum sambac : plant in flowering stage

Jasminum sambac : fruits

Khair (*Acacia catechu*)

Commonname : Catechu tree

Family : Fabaceae sab f. Mimosoideae

According to Vamana Purana the khadir tree arose from the middle body of Brahma. In Hindu mythology, khair is a sacred tree and associated to kartikeya

Acacia catechu : plant in flowering stage

Acacia catechu : flowers

planet Mars.This tree is also sacred in Buddhism. Its wood is used in the religious ceremonies at the time of havans or homa. Khair wood also used at the funeral ceremony. It is said that khair provide moksha to the diseased soul. Kattha is used with pan supari in various rituals in Uttarakhand.

Mehandi (*Lawsonia inermis*)

Common name : Henna

Family : Lythraceae

In Hindu religion mehandi is a symbol of festivity and most auspicious ornament of married women. Women also applied mehandi during various fasts like Karvachauth. Mehandi also be seen on the palms of various God and Goddess. It is used in treatment of different ailments amoebic syscentry, diarrhoea, throat sores etc. The leaves are used by women for dying skin, hair, clothes and nails whereas men use for coloring hair and beard.

Lawsonia inermis : plant in flowering stage

Lawsonia inermis : fruits

6

Major Tribes of India and Thier Ethnobotany

India, the land of oldest civilization known through out the world for its vibrant cultures, faiths, lifestyle, traditions, rituals etc. which are interwoven with each other since time immemoral. Here people of different ethnic groups live together who speak different languages and dialects according to their religion and choice. Altogether, 68 million inhabitants belonging to 227 ethnic groups, comprising of 573 tribal communities exists in different part of the country. These people constitute 8.08% of the total population of India (1991 Census of India) and speaks <750 dialects that can be classified into one of the following three language families: Austro-Asiatic (AA), Dravidian (DR), and Tibeto-Burman (TB). When come to think of different tribal community's inhabitating India, they can be easily divided into six major racial lineages i.e. - Negroid, Mongoloid, Proto-Australoid/ Austrics, Mediterranean or Dravidian, West Brachycephals and Nordic Aryans (Fig. 6.1).

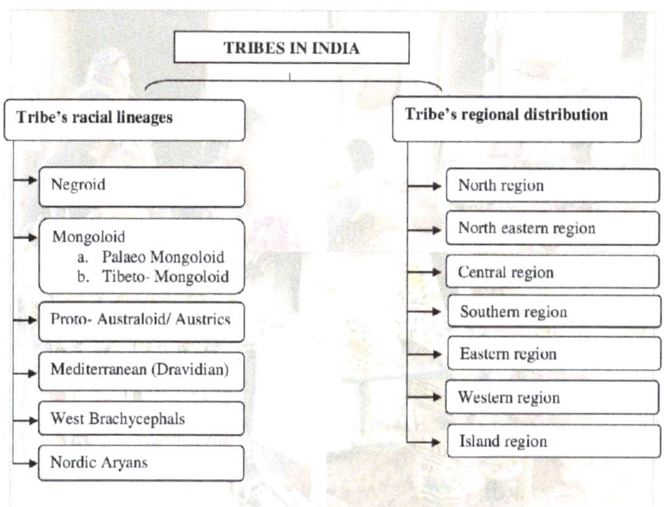

Fig. 6.1. Racial lineages & regional Distribution of Tribes in India.

However, for a better understanding of the tribal region segregation, India can be divided into seven zones i.e. North zones, North-eastern zones, Central zones, Southern zones, Eastern zones, Western zones and Island (Fig. 6.2). In these zones these tribal people survive in close proximity of the forests and manage and conserve biodiversity near their surroundings since ages. These people utilize forest produce, timber and fuelwood and countless of additional raw materials from forest and make use of number of wild medicinal plants for curing their various ailments.

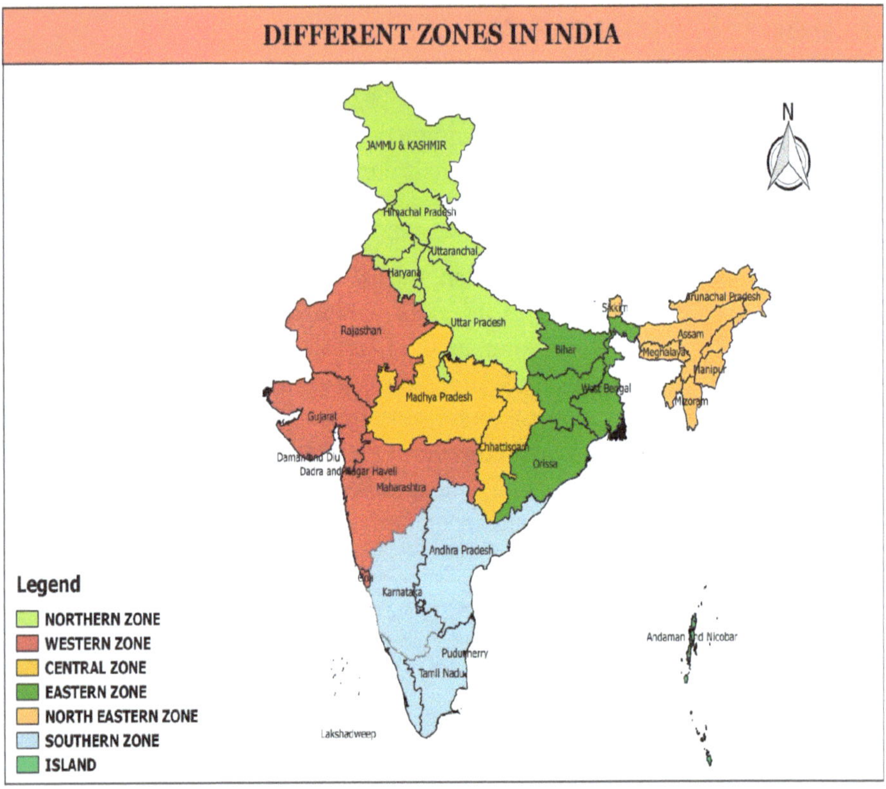

Fig. 6.2

In term of biodiversity, Indian is one of the twelve mega biodiversity centres having two biological diversity hotspot one occurring in Western Ghat and another one in North-Eastern Himalaya. Floristically, 141 endemic genera belonging to over 47 families of higher plant occur in India. Botanical survey of India has reported 46,214 plant species are found in India of global flora of these 17,500 represents flowering plants. Thirty seven of these are endemic and found in North-East of India. In India 11.95% of the world's biodiversity has been conserved by ethnic people in many ways. Out of total reported wild plant in India i.e. 45,000 species;approx 9,500 species are ethnobotanically important and of which 7,500 plant species are medicinally known in traditional health care systems. Approximate 3,900 plant species are used by natives as food (out of which 145 species comprise of root and tubers, 521 species of leafy vegetables, 101 species of bulbs and flowers, 647 species of fruits), 525 species are used as fiber, 400 species as fodder, 300 species are used in preparation and extraction of bio-chemicals which are used as naturally occurring insecticides and pesticides, 300 species are used for extraction of gum, resins, dyes and perfume. Besides this a lots many plants are used as timber, for building material and about 700 species are culturally important from moral, cultural, religious, aesthetic and social point of view.

Since it is known that the tribal people are not only familiar with the knowledge of plant species in their traditional health practices, but also understand the ecological interactions of the various components of their resources. Therefore, ethno-medico-botany of wide variety of plants use by major tribes residing in different zone of India is discussed in this chapter. This will create awareness amongst young generation and motivate them to acqure this traditional medicinal knowledge and help the society in conservation of this valuable traditional knowledge of our nation.

Zonal distributions of tribes and their ethnobotanical practices

1. North Zone

In this zone tribes Jammu & Kashmir, Himachal Pradesh, Punjab, Uttarakhand, Sub Himalayan Uttar Pradesh and Bihar comes i.e. Gujjars, Lahul, Lepcha, Bhotia, Tharu, Buxa, Jaunsari, Khampa, Bhoksa, Gaddi, Pangal and Kanauta. They all have the characteristics of the Mongoloids racial group (Fig. 6.3).

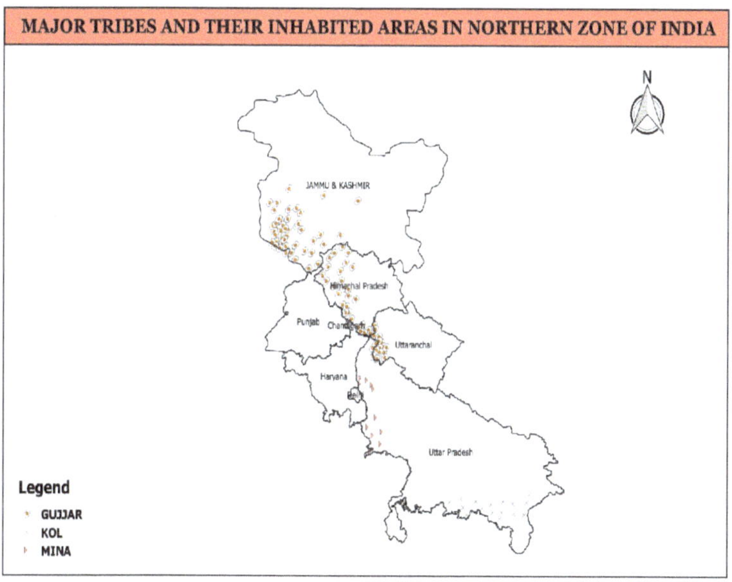

Fig. 6.3

Several forest fringe communities inhabiting in North India particularly in the rural areas are dependent on the plant derived medicines which are prepared from locally available plants. The details of few priority species use to cure different ailments in Northern region of India are given below:

Table 6.1. Priority species used for traditional healing practices in North zone of India.

S. No.	Botanical name of plant	Vernacular name	Family	Plant part/s used	Uses
1.	*Abelmoschus moschatus*	Lata kasturi	Malvaceae	Seeds	Carminative
2.	*Abies pindrow*	Sal	Pinaceae	Bark	Rheumatism
3.	*Achillea millefolium*	Berguer, Pahal gassesh	Asteraceae	Rhizome, Leaves	Headache, Cough, Tooth ache
4.	*Achyranthes aspera*	Chirchita	Amaranthaceae	Seeds	Fever
5.	*Aconitum heterophyllum*	Paewakh	Ranunculaceae	Root	Antidote for snake bites,
6.	*Acorus calamus*	Via-gander	Acoraceae	Rhizome	Stomachic, Diarrhea, cough, Swellings, Joint pain
7.	*Ajuga integrifolia*	Neelkanthi	Lamiaceae	Leaves	Diuretic, febrifuge and tonic
8.	*Alstonia scholaris*	Satparni	Apocynaceae	Leaves	Levaer tonic
9.	*Andrographis paniculata*	Kalmegh	Acanthaceae	Whole plant	Antipyretic
10.	*Anemone obtusiloba*	Srub	Ranunculaceae	Seeds	Rheumatism
11.	*Aquilegia fragrans*	Daduejaid	Ranunculaceae	Flowers	Indigestion
12.	*Arisaema jacquemontiana*	Hapat makei		Rhizome	Muscular strength and Skin
13.	*Arnebia benthamii*	Kah Zaban	Boraginaceae	Rhizome	Common Cold, Cough, Fever, blood purifier
14.	*Artemesia absenthium*	Tethwan	Asteraceae	Leaves	Obesity, Diabetes, liver infection
15.	*Atropa acumniata*	Chella lubbar	Solanaceae	Roots and leaves	Cough. And antispasmodic
16.	*Berberis asiatica*	Kingoda	Berberidaceaeae	Roots	Diabetes
17.	*Berberis lyceum*	Kawdach	Beriberidaceae	Roots, Fresh fruit	Indigestion, Constipation.
18.	*Bergenia ciliata*	Pathar chatta	Saxifragaceae	Leaves as well as whole plant	Dissolving kidney stones
19.	*Berginia ligulata*	Zakhmi hayat	Saxifraceae	Leaves and roots	Intestine complaints and stomach ulcers
20.	*Betula utilis*	Bhuz	Betulaceae	Bark	Antiseptic
21.	*Boehmeria rugulosa*	Genti	Urticaceae	Bark	Blood coagulation
22.	*Boenninghausenia albiflora*	Pissumar Booti	Rutaceae	Leaves, Roots, Whole plant	Treatment of scabies, malaria, Relieve headaches

23.	*Boerhavia diffusa*	Bishkhap	Nyctaginaceae	Leaves and stem	Piles, urinary troubles, dropsy condition, Inflammatory renal diseases
24.	*Cannabis sativa*	Bhang	Cannabinnaceae	Leaves, seeds and stem	ear-ache, blood purifier, scabies and piles
25.	*Cardamine impatiens*	Pahal-laish	Brassicaceae	Whole plant	Asthma, Hay fever
26.	*Cascuta reflexa*	Kukliporte	Cuscutaceae	Whole Plant	Joint pains, wound healing and falling of Hairs
27.	*Celastrus paniculatus*	Papkakani	Celastraceae	Seeds	Seed oil applied to skin itching
28.	*Cichorium intybus*	Kazal-Handh	Asteraceae	Root	Rheumatism Sore throat, jaundice
29.	*Colebrookea oppositifolia*	Binda	Lamiaceae	Leaves	Wounds healing
30.	*Commelina benghalensis*	Kansura	Commelinaceae	Leaves	Swelling in body
31.	*Coriandrum sativum*	Danival	Apiaceae	Seeds	Hair fall
32.	*Cotula anthemoids*	Bobul	Asteraceae	Roots	Constipation
33.	*Curcuma zeodaria Roscoe*	Kachoor	Zingiberaceae	Rhizomes	Paste applied on body on wedding ceremony
34.	*Cynodon dactylon*	Dubla	Poaceae	Roots	Fever
35.	*Dicliptera bupleuroides*	Kuthhi	Acanthaceae	Seeds and leaves	Used for dysentery in children
36.	*Dioscorea bulbifera*	Genthi	Dioscoreaceae	Fruits	Diabetes
37.	*Dioscorea deltoidea*	Kraeth	Discoreaceae	Leaf	Opthalimic infections, urinary infections
38.	*Diplazium esculentum*	Lingoda	Athyriaceae	Leaves	As vegetable
39.	*Elephantopus scaber*	Neelkanthi	Asteraceae	Leaves	Cure for intestinal worms specially for kids
40.	*Eryngium foetidum*	Phari dhania	Apiaceae	Roots	Stomatic
41.	*Eupatorium adenophorum*	Kharna	Asteraceae	Leaves	Wounds
42.	*Euphorbia helioscopa*	Gurisochol, Gandi booti	Euphorb-iaceae	Seeds, roots and latex	abdominal cramps, cholera and eruptions
43.	*Euphorbia wallichia*	Guri-dud/ Harbi	Euphorbiaceae	Stem, leaves, latex	Skin diseases, and asthma
44.	*Fagopyrum acutatum*	Ban ogal, Kanjolya	Polygonaceae	Whole plant	Anodyne, anthelmintic, carminative, depurative and febrifuge
45.	*Ficus auricalata*	Timla	Moraceae	Fruits	Quickly heal clot
46.	*Ficus carica*	Anjeer	Moraceae	Stem, milkylatex, fruit pulp	Birthrate control,insect bite and warts.
47.	*Ficus palmata*	Bedu	Malvaceae	Fruits	As fruit
48.	*Fumaria indica*	Pugsley, Shahtaur	Fumariaceae plant	Whole	Dyspepsia, Rheumatism

49.	*Grevillea Robusta*	Silver oat	Proteaceae	Roots	Diuretic spamolytic
50.	*Grewia optiva*	Bhimal	Moraceae	Leaves, Branches	Hair dandruff and anti fungal as well as fine fiber
51.	*Hedychium spicatum*	Kapurkachri	Zingiberaceae	Rhizome, leaves	Aromatic, anti-arthritic, appetizer, cardiac stimulant, carminative, deodorant, hair tonic and stomachic
52.	*Hyoscyamus niger*	Bazarbang	Solanaceae	Seed	Tooth ache
53.	*Indigofera heterantha*	Jandi	Leguminosae	Leaves	Internal body disorders
54.	*Iris kashmiriana*	Mazarmund	Iridaceae	Whole plant	Joint pains
55.	*Jatropha curcas*	Rattan jot	Euphorbiacea	Stem bark	Scurvey
56.	*Juglans regia Linn.*	Akhrot	Juglandaceae	Fruits	Cure for ringworm
57.	*Juniperus communis*	Bithur	Curpessaceae	Leaves	rheumatism
58.	*Lavetera kashmeriana*	Sozposh	Malvaceae	Flower	Mumps, skin irritation in pregnant women
59.	*Malaxis acuminata*	Jeevak	Orchidaceae	Pseudobulbs	Fever, seminal weakness, burning sensations, tuberculosis and general debilitY
60.	*Malva sylvestris*	Sotal	Malvaceae	Seeds	Cough, fever,eye sight
61.	*Mesua ferrea*	Nagkesar	Calophyllacea	Flower	Bleeding Piles
62.	*Morinda oleifera*	Sajan	Moringaceae	Stem bark	Indigestion
63.	*Nasturtium officinalle*	Kulhak	Brasicaceae	Leaf	Stomachic
64.	*Nicotiana tabacum*	Tambakhu	Solanaceae	Leaves	Cough
65.	*Ocimum gratissimum*	Rama tulsi	Lamiaceae	Leaves	Cough
66.	*Oxalis corniculata*	Tsok-tsen	Oxalidaceae	Whole plant, leaves	Toothache, Convulsions, Blood purification, Diarrhoea
67.	*Oxalis corniculata*	Bhilmori	Oxalidaceae	Leaves	Cough and bronchitis
68.	*Papaver somniferum*	Kashkhas	Papaveraceae	Fruit	Dry cough, Diarrhea
69.	*Phytolacca acinosa*	Brand	Phytolaccaceae	Root	narcotic effect, sedative
70.	*Pinus roxburghii*	Chad	Pinaceae	Seeds and gums	General weakness after child birth child
71.	*Podophyllum hexandrum*	Banwangun	Berberidaceae	Leaves and roots	Skin diseases, Gastric problems
72.	*Pyrus pashia*	Mehal	Rosaceae	Fruits	Digestive disorder
73.	*Rheum emodi*	Pambechalan	Polygonaceae	Leaves	Rheumatic pain, Wounds, Dislocated joints, Boils

74.	*Rhodiola himalensis*	Dand jari	Crassulaceae	bark	infection of teeth
75.	*Rhus parviflora Roxb.*	Saunla	Anacardiaceae	Branches	Teeth cleaner
76.	*Rosa webbiana*	Gulab	Rosaceae	Flowers	Cough and colds.
77.	*Roylea cinerea*	Kadhwi	Lamiaceae	Leaves	Used in heal the boil
78.	*Rubia cordifolia*	Rubes	Rubiaceae	Roots	Stomachache, Jaundice
79.	*Rubia cordifolia*	Majit, Manjit, Manjhist	Rubiaceae	Whole plant	Antidote to scorpion sting, has antiseptic, astringent, deobstruent, diuretic and vermifuge properties
80.	*Rubus ellipticus*	Hinsar	Rosaceae	Whole plant	Diarrhea
81.	*Rumex hastatus*	Almoru	Polygonaceae	Leaves	Use for healing on wounds and cuts
82.	*Salix wallichiana*	Danthiveer	Salicaceae	Leaves	Fever, Head ache, Genral body pain body
83.	*Sambucus wightiana*	Kown	Sambucaceae	roots, leaves and berries	Diuretic, purgative
84.	*Sapindus mukorossi*	Reetha	Sapindaceae	Fruits	Hair shampoo
85.	*Sigesbeckia orientalis*	Lichkur	Asteraceae	Leaves	For heal wounds & blood clots
86.	*Stellaria media*	Losdhi	Caryophyllaceae	Seed	Skin infection, allergy
87.	*Stephania glabra*	Gindaru	Menispermaceae	Roots	Stomachache, constipation
88.	*Swertia paniculata*	Charaita	Gentianaceae	Whole plant	Treatment of malaria and other fever
89.	*Tagetes minuta* L.	Gaynda	Asteraceae	Leaves	Antiseptic
90.	*Taraxacum officinale*	Hand	Asteraceae	Roots	Back pain, common cold, Chestinfection
91.	*Thalictrum clavatum*	Kirmoli	Ranunculaceae	Roots	Cured worms in to kill intestine worms of the kid
92.	*Thalictrum foliolosum*	Mamira, Mamiri	Ranunculaceae	Roots	diuretic, purgative, and bitter tonic during convalescence and dyspepsia
93.	*Toona ciliata*	Toon	Meliaceae	Bark	Healing wounds as well as timber
94.	*Tribulus terrestris*	Gokharu	Zygophyllaceae	Seeds	Gout and urinary disorder
95.	*Trigonella foenum-graecum*	Meth	Fabaceae	Seeds	Back Pain
96.	*Urtica dioca*	Soi	Urticaceae	Leaves and Roots	Rheumatism

97.	*Urtica dioica*	Kandali	Urticaceae	Whole plants	Stem used as fiber; fresh leaves as vegetable
98.	*Valeriana jatamansi*	Samoya	Valerianaceae	Whole Plant	Leaves kept in clothes, paste applied on body on wedding
99.	*Viburnum grandiflorum*	kulmanch	Caprifoliaceae	Seed	Typhoid, whooping cough
100.	*Viola odorata*	Bunufsha	Violaceae	Leaves, seeds and flowers	Respiratory problems
101.	*Viscum album*	Aal	Loranthaceae	Whole plant	Laxative and Fractures
102.	*Zanthoxylum armatum*	Timru	Rutaceae	Stem	Mouth cleaner

2. North-Eastern zone

North-east zone is part of both "Himalaya" and "Indo-Burma" *biodiversity* hotspots. It comprising the states of Arunachal Pradesh, Assam, Manipur, Meghalaya, Mizoram, Nagaland, Tripura and Sikkim lies between 21°34'N to 29°50'N latitude and 87°32'E to 97°52'E longitudes and covers an area of 262060 sq. km. The region harbors nearly 50% of the flowering plants recorded from India and exhibits the richest diversity due to its diversified topography and climatic conditions. The region is home to many wild relative of cultivated plants such as Orchids, Musa, Bamboos, Citrus, Gingers, Palms, etc. More than 250 tribes of different ethnic (Fig. 6.4). The details of imporatant species use to cure different ailments by major tribes of Northern eastern zone of India are given below:

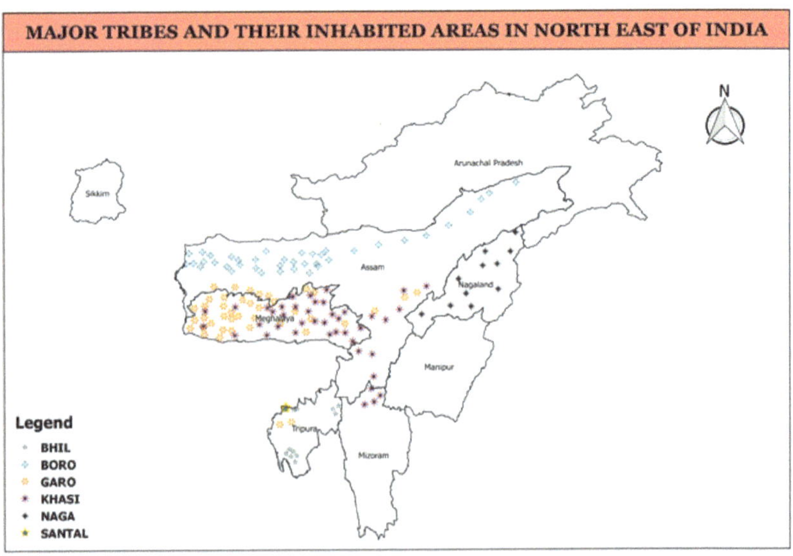

Fig. 6.4

These tribes of North Eastern region have been categorized into two broad ethnic communities, such as the Khasi and the Jaintia tribe of Meghalaya, who belong to 'Monkhemar' culture of Austoic dialect and the rest of the tribal groups are basically Mongoloid, who belongs to Tibeto-Burman sub family of Tibeto-Chinese group (Dutta & Dutta, 2005). Predominantly the bulk of these ethnic communities are forest dwellers that have accumulated a rich knowledge on the uses of various forests and forest products over the centuries. The details of major tribes inhabiting North-East India and their ethnobotanical practices are given below:

Table 6.2 : Major ethnic communities of North-Eastern zone of India and their ethnobotanical practices

S. No.	Scientific name	Local name	Uses
Arunachal Pradesh			
Tribes: Dafla, Khampti and Singpho (Source: Deb & Sharma 2015; Dutta & Dutta 2005).			
1.	*Abroma augusta* L.	Pishach Karpasa	Root bark is used in leucorrhoea.
2.	Amaranthus spinusus Lin	Gobrai	Antidote in snake bite; gonorrhea
3.	*Bryophyllum calycinum Sailisb*	Yapong	Urinary Troubles
4.	*Bulmea fistulosa* Kurz	Rumbdum	Diarrhea
5.	*Calotropis gigantean* (L.) R.Br. ex. Ait.	Kawenexait	Used in sexual weakness
6.	*Cassia alata* L.	Dadmardan	Bronchitis and asthma and against ring worm
7.	*Centella asiatica* (L.) Urban	Brahmi	Arthritis, diabetes, blood disorders and brain tonic.
8.	Coffca bengalensis Roxb	Wanco	Stomach disorder
9.	*Coptis teeta* Wall.	Mishmi teeta	Malaria, jaundice, diabetes
10.	*Drymaria cordata* Willd.	Mani-moone	Laxative
11.	*Elaeocarpus floribundus* Blume	Jalpai	As mouth wash for inflamed gums
12.	*Litsaea cubeba* (Lour.) Pers.	Santero	Cold and cough; Delivery trouble
13.	*Mesua ferrea L.*	Naharpatiya	Used on wounds
14.	*Mussenda roxburghii* Hook. F	Aksap	Malaria, jaundice
15.	*Oxalis corniculate* Linn	Phakep	Relieves intoxication from wine; Diarrhea
16.	Piper attaenuatumBuch.-Ham. ex Miq.	Dolopan	Liver and urinary troubles
17.	*Solanum spirale Roxb.*		Hypertension, sexual diseases
18.	*Solanum viarum* Dunal.	Koppir	Contraceptive properties
19.	*Spilanthes acmella* Murr.	Marshang	Tooth ache
20.	*Swertia chirayita* (Roxb. ex Flem.) Karst.	Chirata	Malaria, fever
21.	*Tamarindus indicus L.*	Tentul	Used for conjunctivitis
22.	*Z. hamiltonianum* Wall.	Ombeng	Malaria
23.	*Zanthoxylum alatum* Roxb.	Timbru	Digestion, tooth ache

Assam

Tribes: Karbi, Dimasa, Mikirs, Jaintia, Rabha, Deori & Mishi

(Source: Singh *et al.* 2017 & 2018; Deb *et al.* 2015)

24.	*Asparagus racemosus* Willd.	Satmul	Urinary disorder
25.	*Averrhoa carambola* L.	Kardoi	Jaundice, diabetes
26.	*Bombax ceiba L.*	Semul	Control diabetes
27.	*Bryophyllum calycinum* (Lamarck) Oken	Patharchura	Urinary disorder
28.	*Costus speciosus* (Koe.) Sm.	Debitokan	Dysurea
29.	*Cleome gynandra* L.	Bhutmulla	Inflammatory diseases
30.	*Dioscorea alata L.*	Kath alu	Anthelmintic, treat leprosy, piles
31.	*Emblica officinalis* Gaertn.	Amloki	Anti-inflammatory, antiulcer, hepatoprotective
32.	*Euphorbia neriifolia* L.	Siju	Cough
33.	*Garcinia tinctoria* (Dc.) Wight	Cuithekera	Rheumatism and bowel complaints
34.	Gnetum montanam Markgr.	Mameilet	Malaria, antidote to poisons
35.	*Justicia adhatoda* L.	Tita bahak	Cough
36.	*Leucas aspera* Spreng.	Takom Kori	Indigestion
37.	*Melia azadarachta* L.	Mahanium	Cough
38.	*Musa balbisiana* Colla.	Kopak, kola	Cough
39.	*Nymphaea nouchali Burm.*	Kumud	Indigestion, flatulence
40.	*Ocimum sanctum* Linn.	Tulosi	Cough, curing allergy
41.	*Oroxylum indicum* (L.) Bent.	Ghilayesing	diabetes
42.	*Plumeria acutifolia* Por	Golancha	Antifertility
43.	*Polygalachinensis* L.	Meradu	Asthma, fever and dizziness
44.	*Portulaca oleracea* L	Malbhog xaak	cardio-vascular disease, dysuria, haematuria, sore nipples and mouth ulcer
45.	*Randia spinosa* Thunb.	Gurol	skin diseases, anti-inflammatory
46.	*Rauwolfia serpentina* Benth. Ex Kurz.	Arachoritita	Intestinal troubles, stimulate uterine contraction
47.	*Sida cordifolia* L.	Bala	Jaundice
48.	*Swertia chirayita* Sw.	Chiraita	Malaria, fever
49.	*Tamarix dioica*		
50.	*Vetiveria zizanioides* L.	Birina	Gastrointestinal disorders
51.	*Vitex negundo* Linn.	Pochotia	eczema

Manipur

Tribes: Meities, Pangals, Naga tribes and Kuki (Source: Deb *et al.* 2015)

52.	*Acorus calamus* L.	Ok-Hidak	Cough and chest congestion; dysentery
53.	*Aquilaria agallocha* Roxb.	Agar	Astringent in diarrhoea and vomiting; snake bite.
54.	*Baccaurea ramiflora* Lour.	Leteku	Constipation
55.	*Clerodendrum infortunatum* L.	Kuthap	Pidika (boils)
56.	*Costus speciosus* Sm.	Khongbam	Urinary complaints, snake bite
57.	*Goniothalamus sesquipedalis* Hook. F. &Thoms	Leikham	Bath for new born child
58.	*Kaempferia rotunda* L.	Yai Thamna	Mumps, tumors, swelling and wounds
59.	*Lemania australis*	Nungsham	Abortifacient drug
60.	*Oxalis corniculata* L.	Yensil	Abdominal pain
61.	*Plectranthus ternifolius* D. Don	Khoiju	Hair care lotion, antidote of small pox, skin diseases
62.	*Blumea balsamifera* (Linn.) DC.	Langthrei	Blood pressure
63.	Rhododendron arboreum Sm.	Khorom-leishak-	Reduce blood pressure
64.	*Andrographis paniculata* Nees	Kalmegh	Leucorrhoea, jaundice, Gonorrhoea

65.	*Artemisia nilagirica* Linn.	Laibak-ngou	Ulcer, dizziness
66.	*Cardamine hirsute* Linn.	Chantruk-maan	Diuretic, better urination
67.	*Gynura nepalensis* D C.	Terapaibi	Stomach ulcer
68.	*Hedychium marginatum* C. B. Clarke	Takhellei-angangba	Bronchitis & stomach ulcer
69.	*Iris bakeri* Chapin.	Kombirei	Brain coolant & hysteria
70.	Stevia rebaudiana Bertoni.	Stevia	Diabetes control
71.	Scutellaria discolor (Wallich. ex Benth.)	Yenakhat	Menstrual pain

Meghalaya

Tribes: Garo, Khasi and Jayantia (Source: Singh *et al.,* 2017; Joseph and Kharkonger 1997)

72.	*Achyranthus bidentata* L.	Minamkachi	Leprosy
73.	*Adiantum phillipense* (Linn.)	Tyrkhang khyllai	Fractured bones.
74.	*Antidesma bunius* L.	Soh-syllai	Joint pains
75.	*Arisaema jacquernontii* Bl.	Jinjok (Garo)	Ringworm and other skin diseses
76.	*Averrhoa carambola* L.	Sohpyrshong	Jaundice
77.	*Betula alnoides* Buch Ham	Dienglieng	Indigestion and flatulence
78.	Dendrobium moschatum (Buch.-Ham.) Sw.	Tiew dieng	Anthelmintic, malarial fever
79.	*Drymaria cordata* (L.) Willd. ex Roem & Schult	Bat-nongrim	Burns, skin diseases, snake bites.
80.	*Elephantopus scaber* L.	Achaksn (Garo).	Heart and liver trouble
81.	*Eleusine indica* (L.)	Lang krai	Jaundice
82.	*Gmelina arborea* L.	Gomari	Cough, gonorrhoea
83.	*Gynocardia odorata* R.Br	Sohliang	Leprosy, nausea and rheumatism
84.	*Hedyotis diffusa Willd.*	Mo-shoh shu	Gastric troubles, cough, cold
85.	*Hedyotis scandens* Roxb.	Bhadeli-lota	Cough and cold
86.	*Holarrhena antidysenterica* (L.) Wall.	Minamkachi	Leprosy
87.	*Kaempferia rotunda* L.	Ingsmoh	Stomach trouble
88.	*Litsea cubeba* (Lours.) Pers.	Jinjok	Ringworms; Also applied for various skin diseases
89.	*Melia azedarach* L.	Dieng ja rasang	Anthelmintic
90.	*Nepenthes khasiana* Hk.f.	Ksetphare	Stomach-ache, eye sores or urinary troubles.
91.	*Osbeckia stellata* Buch.Ham ex D. Don	Soh-lyngkthut	Against snake bites, bleeding
92.	*Oxalis corniculata* (L.	Jabuit	Diarrhoea
93.	*Poulzolzia hirta* Hassk.	Chiple ghaans	Boil, bone setting
94.	*Pouzolzia hirta* Bl.	Memsleh	Hair tonic
95.	*Rauvolfia serpentina* Benth. ex. Kurz.	Sarpagandha	Antidote ; increase uterine contraction; bowels; blood pressure
96.	*Rubia cordifolia* L.	Bad-rahoi (Khasi).	Ulcers
97.	*Solanum viarum* Dunal.	Toothache	

Table 2 : Major ethnic communities of North-Eastern zone of India and their ethnobotanical practices

S No.	Scientific name	Local name	Uses
Arunachal Pradesh			
Tribes: Dafla, Khampti and Singpho (Source: Deb & Sharma 2015; Dutta & Dutta 2005).			
98.	*Abroma augusta* L.	Pishach Karpasa	Root bark is used in leucorrhoea.
99.	*Amaranthus spinusus* Lin	Gobrai	Antidote in snake bite; gonorrhea
100.	*Bryophyllum calycinum* Sailisb	Yapong	Urinary Troubles
101.	*Bulmea fistulosa* Kurz	Rumbdum	Diarrhea
102.	*Calotropis gigantean* (L.) R.Br. ex. Ait.	Kawenexait	Used in sexual weakness
103.	*Cassia alata* L.	Dadmardan	Bronchitis and asthma and against ring worm
104.	*Centella asiatica* (L.) Urban	Mandookaparni/ Brahmi	Arthritis, diabetes, blood disorders and brain tonic.
105.	*Coffca bengalensis* Roxb	Wanco	Stomach disorder
106.	*Coptis teeta* Wall.	Mishmi teeta	Malaria, jaundice, diabetes
107.	*Drymaria cordata* Willd.	Mani-moone	Laxative
108.	*Elaeocarpus floribundus* Blume	Jalpai	As mouth wash for inflamed gums
109.	*Litsaea cubeba* (Lour.) Pers.	Santero (Apatani)	Cold and cough; Delivery trouble
110.	*Mesua ferrea L.*	Naharpatiya	Used on wounds
111.	*Mussenda roxburghii* Hook. F	Aksap	Malaria, jaundice
112.	*Oxalis corniculate* Linn	Phakep	Relieves intoxication from wine; Diarrhea
113.	Piper attaenuatumBuch.-Ham. ex Miq.	Dolopan	Liver and urinary troubles
114.	*Solanum spirale* Roxb.	Titakuchi	Hypertension, sexual diseases
115.	*Solanum viarum* Dunal.	Koppir	Contraceptive properties
116.	*Spilanthes acmella* Murr.	Marshang	Tooth ache
117.	*Swertia chirayita* (Roxb. ex Flem.) Karst.	Chirata	Malaria, fever
118.	*Tamarindus indicus* L.	Tentul	Used for conjunctivitis
119.	*Z. hamiltonianum* Wall.	Ombeng (Adi)	Malaria
120.	*Zanthoxylum alatum* Roxb.	Timbru	Digestion, tooth ache
Assam			
Tribes: Karbi, Dimasa, Mikirs, Jaintia, Rabha, Deori & Mishi (Source: Singh *et al.* 2017 & 2018; Deb *et al.* 2015)			
121.	*Asparagus racemosus* Willd.	Satmul	Urinary disorder
122.	*Averrhoa carambola* L.	Kardoi	Jaundice, diabetes
123.	*Bombax ceiba* L.	Semul	Control diabetes
124.	*Bryophyllum calycinum* (Lamarck) Oken	Patharchura	Urinary disorder
125.	*Costus speciosus* (Koe.) Sm.	Debitokan	Dysurea
126.	*Cleome gynandra* L.	Bhutmulla	Inflammatory diseases
127.	*Dioscorea alata* L.	Kath alu	Anthelmintic, treat leprosy, piles
128.	*Emblica officinalis* Gaertn.	Amloki	Anti-inflammatory, antiulcer, hepatoprotective
129.	*Euphorbia neriifolia* L.	Siju	Cough
130.	*Garcinia tinctoria* (Dc.) Wight	Cuithekera	Rheumatism and bowel complaints
131.	Gnetum montanam Markgr.	Mameilet, thanlping-rhui	Malaria, antidote to poisons
132.	*Justicia adhatoda* L.	Tita bahak	Cough

133. *Leucas aspera* Spreng.	Takom Kori	Indigestion
134. *Melia azadarachta* L.	Mahanium	Cough
135. *Musa balbisiana* Colla.	Kopak, kola	Cough
136. *Nymphaea nouchali* Burm.	Kumud	Indigestion, flatulence
137. *Ocimum sanctum* Linn.	Tulosi,	Cough, curing allergy
138. *Oroxylum indicum* (L.) Bent.	Ghilayesing	diabetes
139. *Plumeria acutifolia* Por	Golancha	Antifertility
140. *Polygalachinensis* L.	Meradu	Asthma, fever and dizziness
141. *Portulaca oleracea* L	Malbhog xaak	cardio-vascular disease, dysuria, haematuria, sore nipples and mouth ulcer
142. *Randia spinosa* Thunb.	Gurol, behmona	Skin diseases, anti-inflammatory
143. *Rauwolfia serpentina* Benth. Ex Kurz.	Arachoritita	Intestinal troubles, stimulate uterine contraction
144. *Sida cordifolia* L.	Bala	Jaundice
145. *Swertia chirayita* Sw.	Chiraita	Malaria, fever
146. *Vetiveria zizanioides* L.	Birina	Gastrointestinal disorders
147. *Vitex negundo* Linn.	Pochotia	Eczema

Manipur

Tribes: Meities, Pangals, Naga tribes and Kuki (Source: Deb *et al.* 2015)

148. *Acorus calamus* L.	Ok-Hidak	Cough and chest congestion; dysentery
149. *Aquilaria agallocha* Roxb.	Agar	Astringent in diarrhoea and vomiting; snake bite.
150. *Baccaurea ramiflora* Lour.	Leteku	Constipation
151. *Clerodendrum infortunatum* L.	Kuthap	Pidika (boils)
152. *Costus speciosus* Sm.	Khongbam takhelei	Urinary complaints, snake bite
153. *Goniothalamus sesquipedalis* Hook. F. &Thoms	Leikham	Bath for new born child
154. *Kaempferia rotunda* L.	Yai Thamna Manbi	Mumps, tumors, swelling and wounds
155. *Lemania australis*	Nungsham	Abortifacient drug
156. *Oxalis corniculata* L.	Yensil	Abdominal pain
157. *Plectranthus ternifolius* D. Don	Khoiju	Hair care lotion, antidote of small pox,skin diseases
158. *Blumea balsamifera* (Linn.) DC.	Langthrei	Blood pressure
159. Rhododendron arboreum Sm.	Khorom-leishak-angangb	Reduce blood pressure
160. *Andrographis paniculata* Nees		Leucorrhoea, menorrhagi, malaria, jaundice, gonorrhoea
161. *Artemisia nilagirica* Linn.	Laibak-ngou	Ulcer, dizziness
162. *Cardamine hirsute* Linn.	Chantruk-maan	Diuretic, better urination
163. *Gynura nepalensis* D C.	Terapaibi	Stomach ulcer
164. *Hedychium marginatum* C. B. Clarke	Takhellei-angangba	Bronchitis & stomach ulcer
165. *Iris bakeri* Chapin.	Kombirei	Brain coolant & hysteria
166. Stevia rebaudiana Bertoni.	Stevia	Diabetes control
167. *Scutellaria discolor* (Wallich. ex Benth.)	Yenakhat	Menstrual pain

Meghalaya

Tribes: Garo, Khasi and Jayantia (Source: Singh *et al.* 2017; Joseph and Kharkonger 1997)

168. *Achyranthus bidentata* L.	Minamkachi	Leprosy
169. *Adiantum phillipense* (Linn.)	Tyrkhang khyllai	Fractured bones.
170. *Antidesma bunius* L.	Soh-syllai	Joint pains
171. *Arisaema jacquernontii* Bl.	Jinjok	Ringworm and other skin diseses

172.	*Averrhoa carambola* L.	Sohpyrshong	Jaundice
173.	*Betula alnoides* Buch Ham	Dienglieng	Indigestion and flatulence
174.	Dendrobium moschatum (Buch.-Ham.) Sw.	Tiew dieng	Anthelmintic, malarial fever
175.	*Drymaria cordata* (L.) Willd. ex Roem & Schult	Bat-nongrim	Burns, skin diseases, snake bites.
176.	*Elephantopus scaber* L.	Achaksn	Heart and liver trouble
177.	*Eleusine indica* (L.)	Lang krai	Jaundice
178.	*Gmelina arborea* L.	Gomari	Cough, gonorrhoea
179.	*Gynocardia odorata* R.Br	Sohliang	Leprosy, nausea and rheumatism
180.	*Hedyotis diffusa Willd.*	Mo-shoh shu	Gastric troubles, cough, cold
181.	*Hedyotis scandens* Roxb.	Cough and cold	
182.	*Holarrhena antidysenterica* (L.) Wall.	Minamkachi	Leprosy
183.	*Kaempferia rotunda* L.	Ingsmoh	Stomach trouble
184.	*Litsea cubeba* (Lours.) Pers.	Jinjok	Ringworms; Also applied for various skin diseases
185.	*Melia azedarach* L.	Dieng ja rasang	Anthelmintic
186.	*Nepenthes khasiana* Hk.f.	Ksetphare	Stomach-ache, eye sores or urinary troubles.
187.	*Osbeckia stellata* Buch.Ham ex D. Don	Soh-lyngkthut	Against snake bites, bleeding
188.	*Oxalis corniculata* (L.	Jabuit	Diarrhoea
189.	*Poulzolzia hirta* Hassk.	Boil, bone setting	
190.	*Pouzolzia hirta* Bl.	Memsleh	Hair tonic
191.	*Rauvolfia serpentina* Benth. ex. Kurz.	Sarpagandha	Antidote; increase uterine contraction; bowels; blood pressure
192.	*Rubia cordifolia* L.	Bad-rahoi	Ulcers
193.	*Solanum viarum* Dunal.	Tit-bhekuri	Toothache

Mizoram

Tribes: Lusai, Kuki, Garo, Khasi, Jayantia and Mikir (Source: Deb *et al.* 2015)

1.	*Acampe papilosa* Lindle	Buangban Parbuk	Acute rheumatism, uterine trouble
2.	*Aeschynanthes sikkimensis* (Cl.) Stap.	Bawltehlaitai	Throat pain
3.	*Aquilaria malaccensis.* Lam.	Thingrai	Carminative, diarrhoea and vomitting, snake-bite, constipation.
4.	Bergenia ligulata (Wall.) Engl.	Pandamdawi	Darrhoea and in pulmonary infections
5.	*Buddleia asiatica* Lour.	Sialrial/Serial	Skin diseases
6.	*Butea frondosa* Koen. ex. Roxb	Fartuah or tuahpui	Skin diseases, urinary disorders
7.	*Chickrassia tabularis* Andr. Juss	Zawngtei	Hyperacidity, diarrhoea and fever
8.	*Cinnamomum cassia* (Nees) Nees & Eberm	Thakthing	Dyspepsia and liver complaint
9.	*Cinnamomum glanduliferum* Meissn.	Khianzo	Cough, bronchitis and pneumonia
10.	*Clerodendrum serratum* (L.) Moon	Ram Phuihnam	Malaria.
11.	*Dalbergia pinnata* (Lour.) Prain.	Hruitengtere	Mastifactory and anti-helminthic
12.	*Dillenia indica,* L.	Kawrthindeng	Abdominal pain.
13.	*Dinochloa compactiflora* Kurtz.	Sairil	influenza
14.	*Eugenia jambolana* Lam	Lenhmui	Dysentery, applied externally on sores and ulcers
15.	*Hibiscus surattensis* L.	Leitha	Treatment of irritation of the penis, venereal sores and ureteritis
16.	*Hodgsonia heteroclita* (Roxb.) Hk.f. et. T.	Khaum	uterine disorders

17.	*Impatiens chinensis* L.	Haeirlo	Gonorrhoea
18.	*Lagerstroemia speciosa* (L). Pers.	Chawn-pui	Diarrhoea and dysentery
19.	*Pholidotaarticulata* Lindle	Nauban	Antiseptic
20.	*Smilax lanceaefolia,* Roxb.	Kaiha	Rheumatism
21.	*Vitex peduncularis* Wall.	Thing-khawi-lu	Chest pain

Nagaland

Tribes: Naga, Kuki, Mikir and Garo (Source: Deb *et al.* 2015)

22.	*Clerodendrum glandulosum* Lindle	Nephaphu	Hypertension and malaria
23.	*Curculigo capitulata* (Lour.) Kuntz.	Palm grass	Stomach problem
24.	*Curcuma zedoaria* Rosc.	Tepetila	Body ache & snake bite
25.	*Garcinia paniculata* Roxb.	Sungsulasu	Stomach and Gastric problem
26.	*Houttuynia cordata* Thunb.	Musundari	Germicides
27.	*Lassia spinosa* L.	Chengmora	Gastrointestinal problem
28.	*Nepenthes khasiana* Hook. F.	Shi achiba tong	Cuts and injuries
29.	*Polyalthia longifolia* Benth. et Hk.	Mongmong	Carminative and anthelmintic
30.	*Polygonum hydropiper* L.	Jakremtsu	Fungal infections, itching skin
31.	*Ricinus communis* L.	Era-gach	Used for bone fracture and skin damage
32.	*Sapindus mukorossi*Gaertn .	Ering jang	Foot and toe infection
33.	*Stephenia hernandifolia* (Willd.) Walp.	Takulaizu	Diarrhoea, nausea, vomiting
34.	*Terminalia chebula* Retz.	Harad	Antiemetic, cough suppressor
35.	*Urtica urens* L.	Jaklemtsu	Constipation and stomach disorder
36.	*Verbena officinalis* L.	Shunutamtsu	Fever and malaria
37.	Zingiber officinale Rosc.	Sung sung	Cough, cold, fever and as antidote in food poisoning
38.	*Schima wallichii* (DC.) Korth.	Bangko	Cuts and injuries

Sikkim

Tribes: Lepcha, Bhutia, Limbu and Tamang (Source: Bharti & Sharma 2012; Deb *et al.* 2015)

39.	*Aconitum ferox* Wall.	Bikh, Katbis	Antiperiodic, antidiabitic, antiphologistic, antipyretic, anodyne
40.	*Drymaria cordata* Willd	Avijal	Headache, throat paint pneumonia and dog bites.
41.	*Hedyotis corymbosa* L.	Piriengo	Antispasmodic.
42.	*Heracleum wallichii* DC.	Chimphing	Influenza and sinusitis.
43.	*Hydrocotyle asiatica* Linn.	Golpatta	Reduce high blood pressure
44.	*Knema angustifolia.* Roxb.	Ramguwa	Dysentery
45.	*Nardostachys grandiflora* DC.	Jatamansi	Epillepsy, lowblood pressure
46.	*Nardostachys jatamansi* DC.	Jatamashi	Stop epileptic fits, cholera, and palpitation
47.	*Nephrolepis cordifolia* Presl	Paniawala	Abdominal pain
48.	*Orchis latifolia* Linn.	Salampanja	Treat nervous disorders and as aphrodisiacs
49.	*Picrorhiza kurrooa* Benth.	Kakati	Dysentery
50.	*Piper longum* Linn.	Pipla	Treat respiratory problems
51.	*Rheum emodi* Wall. ex Meissner.	Khokim	Treat fever, body ache, and muscular pain
52.	*Swertia chirayita* (Roxb. ex Flem.) Karst.	Chirata	Fevers, acidity and in biliousdyspepsia accompanied by fever, skin disease
53.	*Taxus wallichiana* Zucc.	Dhengre-salla	Carminative, expectorant, stomachic and tonic.

Tripura

Tribes: Chakma, Garo, Khasi, Kuki, Lusai, Liang and Santhal

54.	*Actinodaphne obovata* Bl.	Pati-khunda	Irregular menstruation

55.	*Amomum lingniforme* (Roxb.) Benth.	Bon Alach	Muscular rheumatism
56.	*Aquilaria malaccensis* Lamk.	Agar	Leucoderma & rheumatism
57.	*Asparagus racemosus* Willd.	Satamuli	Epilepsy
58.	*Erioglossum rubginosum* (Roxb.) Bl.	Muli	Blood dysentery, malarial fever
59.	*Hymenodictyon excelsum* (Roxb.). Wall	Chepkowa	Jaundice, fever
60.	*Jasminum sambac* (Linn.) Ait.	Komali	Wounds healing
61.	*Kaempferia galanga* L.	Homola	Asthma
62.	*Leucas lanata* Benth.	Dron	Cough & cold
63.	*Litsea glutinosa* (Lour.) C.B. Robinson	Kalimanda	Bone fracture or muscle pain
64.	*Melocanna bambusoides* Trin.	Muli bash	Blood coagulation when bitten by leech, mosquito, spider and other insects
65.	*Microcos paniculata* L.	Pitchla	Cure bone fracture
66.	*Musa acuminata* Colla	Ram kala	Curing allergic complains and diabetes
67.	*Pandanus odoratissimus* Linn.	Keya, Ketaki	Cure red urine disorder
68.	*Phajus flavus* (Bl.) Lindl.	Maittehandori	Earache
69.	*Psidium guineense* Swartz.	Jarbogoyam	Scurvy & dentrites
70.	*Terminalia chebula* Retz.	Haritak	Gastric ulcer
71.	*Vernonia cinerea* Less.	Siyalmutra	Cure abdominal pain, gas problems and acute diarrhoea

Mizoram

Tribes: Lusai, Kuki, Garo, Khasi, Jayantia and Mikir (Source: Deb *et al.* 2015)

72.	*Acampe papilosa* Lindle	Buangban Parbuk	Acute rheumatism, uterine trouble
73.	*Aeschynanthes sikkimensis* (Cl.) Stap.	Bawltehlaitai	Throat pain
74.	*Aquilaria malaccensis*. Lam.	Thingrai	Carminative, diarrhoea and vomitting, snake-bite, constipation.
75.	Bergenia ligulata (Wall.) Engl.	Pandamdawi	Darrhoea and in pulmonary infections
76.	*Buddleia asiatica* Lour.	Sialrial/Serial	Skin diseases
77.	*Butea frondosa* Koen. ex. Roxb	Fartuah or tuahpui	Skin diseases, urinary disorders
78.	*Chickrassia tabularis* Andr. Juss	Zawngtei	Hyperacidity, diarrhoea and fever
79.	*Cinnamomum cassia* (Nees) Nees & Eberm	Thakthing	Dyspepsia and liver complaint
80.	*Cinnamomum glanduliferum* Meissn.	Khianzo	Cough, bronchitis and pneumonia
81.	*Clerodendrum serratum* (L.) Moon	Ram Phuihnam	Malaria.
82.	*Dalbergia pinnata* (Lour.) Prain.	Hruitengtere	Mastifactory and anti-helminthic
83.	*Dillenia indica,* L.	Kawrthindeng	Abdominal pain.
84.	*Dinochloa compactiflora* Kurtz.	Sairil	Influenza
85.	*Eugenia jambolana* Lam	Lenhmui	Dysentery, applied externally on sores and ulcers
86.	*Hibiscus surattensis* L.	Leitha	Treatment of irritation of the penis, venereal sores and ureteritis
87.	*Hodgsonia heteroclita*(Roxb.) Hk.f. et. T.	Khaum	uterine disorders
88.	*Impatiens chinensis* L.	Haeirlo	Gonorrhoea
89.	*Lagerstroemia speciosa*(L). Pers.	Chawn-pui	Diarrhoea and dysentery
90.	*Pholidotaarticulata* Lindle	Nauban	Antiseptic
91.	*Smilax lanceaefolia,* Roxb.	Kaiha	Rheumatism
92.	*Vitex peduncularis* Wall.	Thing-khawi-lu	Chest pain

Nagaland

Tribes: Naga, Kuki, Mikir and Garo (Source: Deb *et al.* 2015)

93.	*Clerodendrum glandulosum* Lindle	Nephaphu (Chakma)	Hypertension and malaria
94.	*Curculigo capitulata* (Lour.) Kuntz.		Stomach problem
95.	*Curcuma zedoaria* Rosc.	Kati/Jemsu naro Tepetila (Zel)	Body ache & snake bite
96.	*Garcinia paniculata* Roxb.	Sungsulasu (Ao)	Stomach and Gastric problem
97.	*Houttuynia cordata* Thunb.	Germicides	
98.	*Lassia spinosa* L.	Gastrointestinal problem	
99.	*Nepenthes khasiana* Hook. F.	Shi achiba tong (Ao)	Cuts and injuries
100.	*Polyalthia longifolia* Benth. et Hk.	Mongmong	Carminative and anthelmintic
101.	*Polygonum hydropiper* L.	Jakremtsu	Fungal infections, itching skin
102.	*Ricinus communis* L.		Used for bone fracture and skin damage
103.	*Sapindus mukorossi*Gaertn .	Ering jang	Foot and toe infection
104.	*Stephenia hernandifolia* (Willd.) Walp.	Takulaizu	Diarrhoea, nausea, vomiting
105.	*Terminalia chebula* Retz.	Harad	Antiemetic, cough suppressor
106.	*Urtica urens* L.	Jaklemtsu	Constipation and stomach disorder
107.	*Verbena officinalis* L.	Shunutamtsu	Fever and malaria
108.	Zingiber officinale Rosc.	Sung sung	Cough, cold, fever and as antidote in food poisoning
109.	*Schima wallichii* (DC.) Korth.	Bangko	Cuts and injuries

Sikkim

Tribes: Lepcha, Bhutia, Limbu and Tamang (Source: Bharti & Sharma 2012; Deb *et al.* 2015)

110.	*Aconitum ferox* Wall.	Bikh, Katbis	Antiperiodic, antidiabitic, antiphologistic, antipyretic, anodyne
111.	*Drymaria cordata* Willd	Avijal	Headache, throat paint pneumonia and dog bites.
112.	*Hedyotis corymbosa* L.	Piriengo	Antispasmodic.
113.	*Heracleum wallichii* DC.	Chimphing	Influenza and sinusitis.
114.	*Hydrocotyle asiatica* Linn.	Golpatta	Reduce high blood pressure
115.	*Knema angustifolia.* Roxb.	Ramguwa	Dysentery
116.	*Nardostachys grandiflora* DC.	Jatamansi	Epillepsy, lowblood pressure
117.	*Nardostachys jatamansi* DC.	Jatamashi	Stop epileptic fits, cholera, and palpitation
118.	*Nephrolepis cordifolia* Presl	Paniawala	Abdominal pain
119.	*Orchis latifolia* Linn.	Salampanja or Panchaunlay	Treat nervous disorders and as aphrodisiacs
120.	*Picrorhiza kurrooa* Benth.	Kakati, Katuka	Dysentery
121.	*Piper longum* Linn.	Pipla	Treat respiratory problems
122.	*Rheum emodi* Wall. ex Meissner.	Khokim	Treat fever, body ache, and muscular pain
123.	*Swertia chirayita* (Roxb. ex Flem.) Karst.	Chirata	Fevers, acidity and in biliousdyspepsia accompanied by fever, skin disease
124.	*Taxus wallichiana* Zucc.	Dhengre-salla	Carminative, expectorant, stomachic and tonic.

Tripura

Tribes: Chakma, Garo, Khasi, Kuki, Lusai, Liang and Santhal

125.	*Actinodaphne obovata* Bl.	Pati-khunda	Irregular menstruation
126.	*Amomum lingniforme* (Roxb.) Benth.	Bon Alach (Bhil)	Muscular rheumatism

127.	*Aquilaria malaccensis* Lamk.	Agar (Bhil)	Leucoderma & rheumatism
128.	*Asparagus racemosus* Willd.	Satamuli (Bhil)	Epilepsy
129.	*Erioglossum rubginosum* (Roxb.) Bl.	Muli (Kuki), Aboian (Bhil).	Blood dysentery, malarial fever
130.	*Hymenodictyon excelsum* (Roxb.). Wall	Chepkowa (Kuki)	Jaundice, fever
131.	*Jasminum sambac* (Linn.) Ait.	Komali	Wounds healing
132.	*Kaempferia galanga* L.	Homola (Kuki)	Asthma
133.	*Leucas lanata* Benth.	Dron (Bhil)	Cough & cold
134.	*Litsea glutinosa* (Lour.) C.B. Robinson	Kalimanda	Bone fracture or muscle pain
135.	*Melocanna bambusoides* Trin.	Muli bash	Blood coagulation when bitten by leech, mosquito, spider and other insects
136.	*Microcos paniculata* L.	Pitchla	Cure bone fracture
137.	*Musa acuminata* Colla	Ram kala	Curing allergic complains and diabetes
138.	*Pandanus odoratissimus* Linn.	Keya, Ketaki	Cure red urine disorder
139.	*Phajus flavus* (Bl.) Lindl.	Maittehandori (Kuki)	Earache
140.	*Psidium guineense* Swartz.	Jarbogoyam (Riang)	Scurvy & dentrites
141.	*Terminalia chebula* Retz.	Haritaki (Kuki)	Gastric ulcer
142.	*Vernonia cinerea* Less.	Siyalmutra	Cure abdominal pain, gas problems and acute diarrhoea

3. Central Zone

The tribes of this particular zone are found in the forest areas comprising the states of Madhya Pradesh, Chhattisgarh, Bihar, some parts of Maharashtra and West Bengal. The good examples of the tribes residing in this zone are the **Gonds, Baiga, Maria, Abujhamaria, Bhils, Santhals, and Kols** etc. The central zone of India comprises the highest population of tribes as compare to other zones, with the Gonds considered to be the largest tribe of India. The language of each tribe is unique and their languages spoken are together considered as the austric languages. The tribes of the central zone possess the Proto-Australoid racial features which comprises of dark skinned and well built body, a large forehead etc. (Fig. 6.5).

The Kol tribe of central India particularly Madhya Pradesh and adjoining district Chitrakoot of Uttar Pradesh follows Hindusm. In survy of Manikpur, Chitrkoot it was found that some Kols are also follow Chirstinity and other religion but they all celeberate the religious festivals Dussehra, Holi, Rakhi and Deepawali and usually cremate the dead body apart from the person who is died due to snakebite. Kols are not much interested in performing art but they are nature conservator and depend on forest resources for their livelihood. Their economy is based on collection of Mahua flower and seeds, tendu leaves and collection fuel wood from the forest. They have a rich traditional knowledge about various uses of plants and plant products. Some photographs are given here to show their lifestyle. List of priority species use in traditional healing by tribes of Central zone are given below:

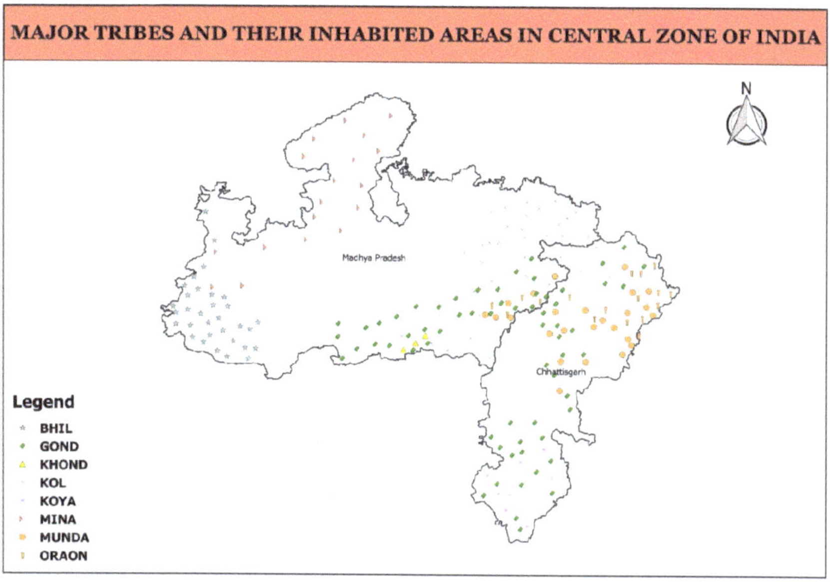

Fig. 6.5

Table 6.3 : Priority species used for traditional healing practices by Baiga and Gond tribes of Central India (Source: Ghosh 2008; Prasad & Singh 2014; Sinha & Chandra, 2011)

S. No.	Species	Local name	Family	Part used	Disease
1.	*Abelmoschus moschatus*	Charmukhi	Malvaceae	Root	Malaria, Itching, Lice, Hair loss, Jaundice,
2.	*Achyranthus aspera*	Chirchita	Amaranthaceae	Root	Asthma, cough
3.	*Adina cordifolia*	Haldu	Rubiaceae	Bark	Wounds, Malaria, Ulcer
4.	*Ageratum conyzoides*	Kubbi	Asteraceae	Leaf	Wound
5.	*Aloe vera*	Guarpatha	Xanthorrhoea-ceae	Leaf	Asthma, Peptic ulcers, Warts, Burns, Pimples
6.	*Alstonia scholaris*	Kuraya	Apocynaceae	Bark	Asthma, tuberculosis, Stomachache, Fever, Cough, Itching
7.	*Amorphophallus paeoniifolius*	Suran, Kanda	Araceae	Rhizome	Rejuvenation, Piles
8.	*Andrographis paniculata*	Kalmegh	Acanthaceae	Leaf	Fever, Malaria, Liver problems
9.	*Anogeissus latifolia*	Dhawa	Combretaceae	Bark	Cough
10.	*Argemone mexicana*	Pili	Papaveraceae	Root	Malaria, Jaundice
11.	*Asparagus recemosa*	Jogilati	Asparagaceae	Tuber	Delivery convalescence, Joint pain, Infertility, Epilepsy

12.	*Azadirachta indica*	Neem	Meliaceae	Roots, Fruits	Malaria, Itching, Lice, Hair loss, Jaundice, Body pain
13.	*Bauhinia vahlii*	Mohlain	Fabaceae	Root	Delivery convalescence
14.	*Begonia picta*	Patharchata	Begoniaceae	Whole plant	Headache, Eye wash
15.	*Biophytum sensitivum*	Lajvanti	Oxalidaceae	Root, Leaf	Itching
16.	*Bombyx ceiba*	Semar	Malvaceae	Root	Weakness
17.	*Boswellia serrata*	Salhey	Burseracaeae	Bark	Cuts especially by iron
18.	*Brassica campestris*	Sarso	Brassicaceae	Seed	Skin disease, Ear cleaning
19.	*Butea monosperma*	Chula	Fabaceae	Seed	Herpes, Skin Disease, Jaundice, Vomiting
20.	*Caesalpinia decapetala*	Kirkich	Fabaceae	Seed	Wounds
21.	*Calotropis procera*	Aakwan	Acanthaceae	Root	Bone Fracture
22.	*Canavalia gladiatum*	Ban Semi	Fabaceae	Root	Body pain, Diarrhoea
23.	*Canna indica*	Bajranti	Cannaceae	Root	Fever, Wounds
24.	*Carissa spinarum*	Karonda	Apocynaceae	Root	Cardiac disorder
25.	*Cassia tora*	Chikoda	Fabaceae	Seeds	Cough
26.	*Catunaregam spinosa*	Mainhar	Rubiaceae	Root	Snake bite
27.	*Centella asiatica*	Brahmi	Apiaceae	Leaf	Memory loss
28.	*Cerisciodes turgida*	Kharhar	Rubiaceae	Root	Snake bite, Joint pain, Muscle problems
29.	*Chlorophytum borivilianum*	Safed musli	Asparagaceae	Tuber	Weakness, Sexual debility
30.	*Cissus quadrangularis*	Hadjod	Vitaceae	Roots	Bone fracture
31.	*Citrus medica*	Attera	Rutaceae	Fruit	Sunstroke
32.	*Colocasia esculenta*	Kochai	Araceae	Rhizome	Asthma
33.	*Cordia macleodii*	Dahiman	Boraginaceae	Bark	Fever, Poisoning, Intoxication
34.	*Costus speciosa*	Kevkand	Zingiberaceae	Rhizome	Weakness
35.	*Cucumis melo*	Indravan	Cucurbitaceae	Fruit, Root	Fever
36.	*Curcuma longa*	Hardi	Zingiberaceae	Root	Cough
37.	*Cynodon dactylon*	Duub Ghas	Poaceae	Leaf	Vomiting
38.	*Cynoglossusm lanceolatum*	Kamraj	Boraginaceae	Root	Eye trouble, Fever
39.	*Datura metal*	Dhatur	Solanaceae	Root	Heat stroke, Genital problem
40.	*Dellinia pentagyna*	Kurkut	Dilleniaceae	Bark	Cuts especially by iron
41.	*Desmodium oojeinense*	Tinsa	Fabaceae	Bark	Bloody stool
42.	*Dioscorea hispida*	Kirchi Kanda	Dioscoreaeceae	Tuber	Ulcer, Wound, Canine and Feline bite
43.	*Dioscorea pentaphylla*	Kaniha Kanda	Dioscoreaeceae	Tuber	Delivery convalescence
44.	*Diospyros melanoxylon*	Tendu	Ebanaceae	Root	Snake bite, Vomiting
45.	*Diospyros montana*	Patwan	Ebanaceae	Bark	Jaundice, Vomiting, Dysentery

46.	*Eclipta alba*	Bhringraj	Asteraceae	Leaf	Wound
47.	*Emblica officinalis*	Amra	Phyllanthaceae	Fruit	Diabetes, fever, Diarrhoea, cough
48.	*Euphorbia hirta*	Dudhiya	Euphorbiaceae	Whole plant	Wound
49.	*Euphorbia hyneana*	Choti Dudhi	Euphorbiaceae	Leaf	Discharge problems
50.	*Ficus religiosa*	Pepar	Moraceae	Prop root	Chest ache
51.	*Ficus virens*	Pakhri	Moraceae	Leaf	Stomach ache
52.	*Flacourtia indica*	Kakai	Salicaceae	Bark	Delivery convalescence
53.	*Gloriosa superba*	Kalihari	Zingiberaceae	Root	Weakness
54.	*Gmelina arbrorea*	Kamher	Lamiaceae	Bark	Respiratory problems
55.	*Grewia hirsuta*	Sakri	Tiliacaeae	Fruits	Weakness
56.	*Helictres isora*	Chota Ainthi	Malvaceae	Fruits	Stomach problems
57.	*Indigofera tinctoria*	Birhul	Fabaceae	Leaf	Headache
58.	*Ipomea balsamina*	Tirraya	Balsaminacae	Leaf	Headache
59.	*Ipomea batatas*	Ratal Kanda	Convulvulaceae	Tuber	Piles
60.	*Ipomea pres-tirgidis*	Khotlaiyan	Convulvulaceae	Seeds	Toothache
61.	*Jatropha curcas*	Ratanjot	Euphorbiaceae	Root	Bloody stool, Diarrhoea
62.	*Leonotis nepetifolia*	Lal Guma	Asteraceae	Infloro-scence	Wound
63.	*Linum usitassimum*	Alsi	Linaceae	Seed	Wound
64.	*Litsea glutinosa*	Maida	Lauraceae	Bark	Boils
65.	*Madhuca indica*	Mahua	Sapotaceae	Fruit, Flower	Delivery convalescence, Skin cracks, Weakness
66.	*Mangifera indica*	Aam	Anacardiaceae	Bark	Jaundice, Headache, Malaria
67.	*Mentha piperata*	Pudina	Apiaceae	Leaf	Indigestion
68.	*Moringa oleifera*	Munga	Moringaceae	Bark, Leaf	Jaundice, Headache
69.	*Mucuna pruriens*	Kewanch	Fabaceae	Root, Seed	Nervous system problems, Sexual debility and Sun stroke
70.	*Murraya paniculata*	Hathil	Rutaceae	Leaf	Diarrhoea
71.	*Musa paradisiaca*	Kela	Musaceae	Leaf	Cough, Diarrhoea
72.	*Opuntia stricta*	Naagfani	Cactaceae	Bark	Boils
73.	*Oxalis corniculata*	Teen Pan	Oxalidaceae	Whole plant	Dyspepsia, Piles, Diarrhoea, Dysentery, Amenhorreae, Hepatitis
74.	*Paspalum scrobiculatum*	Kodo	Poaceae	Seed	Delivery convalescence
75.	*Plumbago zeylanica*	Chita Jadi	Plumbaginaceae	Root	Body pain
76.	*Pongamia pinnata*	Kanji	Fabaceae	Fruit	Itching, Toothache, Rash

77.	*Pueraria tuberosa*	Ban Kumdha	Fabaceae	Root	Paralysis, Weakness
78.	*Punica granatum*	Anar	Lythraceae	Bark	Leucohorrea
79.	*Radermachera xylocarpa*	Garur Phal	Bignoniaceae	Fruit	Skin disease, Abdominal disorder
80.	*Ricinus communis*	Arandi	Euphorbiaceae	Leaf	Burn
81.	*Schleichera oleosa*	Kosum	Sapindaceae	Bark	Stomachache, Bloody stools
82.	*Semecarpus anacradium*	Bhelva	Anacardiaceae	Fruit, Seeds	Headache, Diarrhoea, Paralysis, Worms, Leg ache, Wound, Piles
83.	*Shorea robusta*	Sarai	Dipterocar-paceae	Bark	Delivery convalescence, Diarrhoea, weakness
84.	*Sida acuta*	Kharheta	Malvaceae	Leaf	Headache
85.	*Soymida febrifuga*	Rohina	Meliaceae	Bark	Bleeding, bone fracture, stomachache
86.	*Sterculia urens*	Kurlu	Sterculiaceae	Resin	Sunstroke, Diarrhoea
87.	*Swertia angustifolia*	Hara Chirayta	Acanthaceae	Whole plant	Indigestion, Dyspepsia, Skin disorder, Fever, Abdominal problem,Cancer, Liver problem
88.	*Szygium cumini*	Jamun	Myrtaceae	Bark	Jaundice, Gastric problems, Calculus
89.	*Terminalia alata*	Saja	Combretaceae	Bark	Diarrhoea
90.	*Terminalia arjuna*	Kahuva	Combretaceae	Bark	Weakness
91.	*Terminalia chebula*	Harra	Combretaceae	Fruit	Digestion, Ulcer, Cough, Hiccups, Leprosy, Cardiac disorder, Wound
92.	*Thevetia peruviana*	Pila Kaner	Apocynaceae	Leaf	Itching, Pain
93.	*Tinospora cordifolia*	Padhin	Menispermaceae	Root	Stomachache, Vomiting, Dysentery, Fever, Snake bite, Headache
94.	*Urginea indica*	Van Pyaj	Amaryllidaceae	Tuber	Burn, Swelling, Pregnancy, Vaginalbleeding
95.	*Vetivaria zizanoides*	Khas Ghans	Poaceae	Root	Digestion, Urinary problems, Leprosy
96.	*Woodfordia fruticosa*	Phulchuhi	Lythraceae	Root	Diarrhoea
97.	*Zizhpiphus mautitiana*	Boir	Rhamnaceae	Leaf	Sunstroke, Urinary problem

4. Southern Zone

The tribes residing in thick jungle of Karnataka, western Tamil Nadu, Kerala, western Andhra Pradesh and kerala comes under this zone. The **Toda, Kota, Bagada, Kurumba, Kadar, Paniyan, Chenchu, Allar, Nayak and Chetti** etc are some of the major tribes of this zone. These tribes are known to have the negrito racial features which comprises of dark skinned body, curly and thick hair, well built body with thick lips (Fig. 6.6). Information on priority species and the tribes of Southern zone are given below:

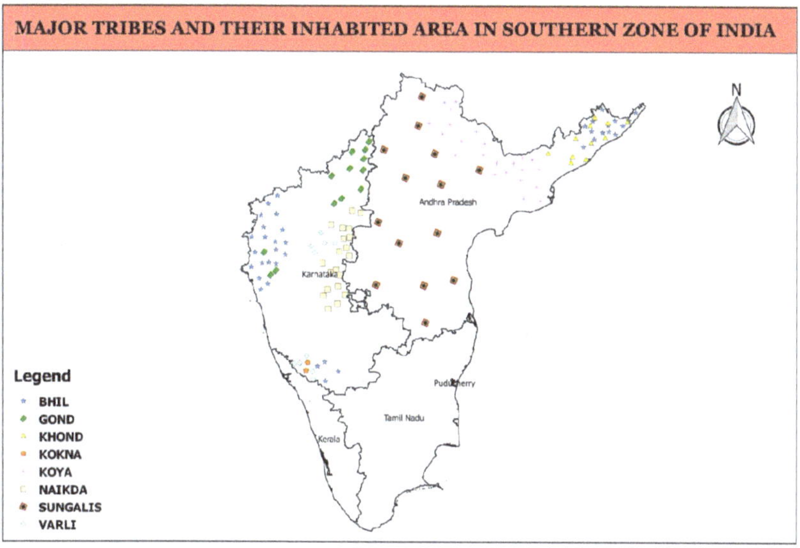

Fig. 6.6

Table 6.4 : Priority species used for traditional healing practices in Southern zone of India

S. No.	Plants Name	Local name	Family	Part used	Medicinal Uses
1	*Acacia catechu* Wild	Kaggali	Fabaceae	Leaves,bark, gum	Skin diseases, fever, herpes, musculoskeletal disorders, dental diseases,
2	*Achyranthes aspera* L	Uttarani	Amaranthaceae	Whole plant	Asthma, itching, hysteria, skin diseases, ulcer, cancer, stone in bladder
3	*Alangium salvifolium* (L.F) Wang	Ankola	Alangiaceae	Leaves, fruit,root	Urinary tract infections,sinus, wounds, snake bite
4	*Aloe vera* (L) Burm.f	Lokasara	Xanthorrho-eaceae	Leaves, pulp	Protecting the skin, burns, sunburn, acne, abdominal tumours

5	*Alternanthera sessilis (L) R. Br.*		Honagone	Amaranthaceae	Whole plant, Acne, asthma, bronchitis
6	*Amaranthus dubius* Mart	Dantu	Amaranthaceae	Leaves, flower, stem	Haemorrhage, kidney problem, anaemia, constipation
7	*Andrographis paniculata* (Burm.f) Wall	Nelabevu	Acanthaceae	Leaves,stem	Malaria, diarrhoea, dysentery, burns, cancer, diabetes, ulcer, leprosy
8	*Argemone mexicana* L	Datturigida	Papavaraceae	Root	Oral ulcers
9	*Azadirachta indica* A. Juss	Bevinamara	Meliaceae	Leaves, bark	Skin problem, dental treatment
10	*Basella alba* L	Baselesoppu	Basellaceae	Leaves, stem	Blood cancer, bruises, burns, anticancer, antiviral
11	*Calotropis gigantean* L	Ekkadagida	Asclepiadaceae	Leaves, roots, latex	Piles, spleen, disorders, antidote for snake poison, cough, skin diseases, diarrhoea, haemorrhage, respiratory problem
12	*Cassia auriculata* L	Avarike	Caesalpinaceae	Leaves, flower, root, seed	Diabetes, fever, urinary diseases conjunctivitis, rheumatism
13	*Cassia tora* L	Chagathegida	Caesalpinaceae	Leaves, seeds	Burns, cuts, wounds, skin problem
14	*Chlorophytum laxum* R.Br	Bicheti grass	Liliaceae	Leaves, herb	
15	*Centella asiatica* (L) Urban	Ondelaga	Apiaceae	Whole plant	Vitamin B, purifies blood, epilepsy, hysteria, agitation, insomnia, urinary tract, infection, asthma
16	Citrus medica L	Madala	Rutaceae	Leaves, seeds, pulp, fruit	Abdominal colic, digestive problem, piles
17	*Coccinia grandis* (L.) Voigt	Thonde	Cucurbitaceae	Leaves, stems	Treatment of diabetes,headache, skin eruptions, chest cold, gonorrhoea, rheumatism
18	*Elaeocarpus ganitrus* Roxb	Rudraksha	Elaeocarpaceae	Leaves, fruit, bark	Nervous system, high blood pressure, tuberculosis, small pox, chicken pox
19	*Emblica officinalis* L	Bettadanelli-kayi	Euphorbiaceae	whole plant	Alopecia, hair treatment, digestive system
20	*Epiphyllum oxypetalum* (DC) Haworth	Brahma kamala	Cactaceae	Flower, stem	Tuberculosis, urinary tract infection, haemorrhage, cough, fever

21	*Ficus racemosa* L	Attimara	Moraceae	Leaves, fruit, root, bark, latex	Leucorrhoea, anaemia, burns, fatigue, leprosy, urinary discharges
22	*Ficus religiosa* L	Aralimara	Moraceae	Bark	Fever, herpes, wounds, bone fracture, skin problem, diarrhoea
23	*Ficus Benghalensis* L	Aaladamara	Moraceae	Bark, aerial root,tender leaf	Fever, fracture, skin diseases, wounds, sinus, diarrhoea
24	*Gymnema sylvestre* R.Br	Madunasini	Apocynaceae	Leaves, seeds, roots	Diabetes, weight loss, stimulating digestive system, diuretic
25	*Justicia adhathoda* L.	Adusoge	Acanthacaeae	Leaves	Used in siddha medicine, Ayurveda Homeopathy and urinary system of medicine
27	*Madhuca longifolia* J.F.Macbr	Hippe	Sapotaceae	Leaves	Skin diseases, nerve disorders, cough, burning sensation
28	*Moringa oleifera* Lam	Nuggemara	Moringaceae	Leaves, fruit, flower, roots	Worms in stomach, iron deficiency, diarrhoea
29	*Murrayakoenigii* L	Karibevu	Rutaceae	Leaves, bark, root	High growth digestion problem, sickness, nausea
30	*Piper betle* L	Veeleyadele	Piperaceae	Leaves, roots	Calcium deficiency, antiseptic, malaria
31	*Plectranthus amboinicus* Spreng	Doddapatre	Lamiaceae	Whole plant	Cold, cough, fever, headache
32	*Pongamia pinnata* (L). Pierr	Hongemara	Fabaceae	Leaves, fruit, bark, twigs	Wounds, skin diseases, piles, musculoskeletal disorders
33	*Portulaca oleracea*L	Doddagoni-soppu	Portulaceae	Leaves, stem, flower	Piles, wound, abdominal disorders, bronchitis, asthma, eye diseases
34	*Pterocarpus marsupium* Roxb	Honnemara	Fabaceae	Leaves, twigs, gum, latex	Fever, piles, skin diseases, musculoskeletal disorders, diabetes, chronic diarrhoea, toothache
35	*Rauwolfia serpentine* L	Sarpagandha	Rutaceae	Leaves, fruits	Bite of poisonous snake, hypertension, low blood pressure
36	*Rubia cardifolia* L	Chitravalli	Rubiaceae	Leaves, fruits, stem, roots,	Skin diseases, arthritis, uric acid, diarrhoea, dysentery ,chronic fever, renal infection, antiseptic diabetes
37	*Saraca asoka* (Roxb) Willd	Ashokamara	Fabaceae	Leaves, bark	Cancer, uterus problems, kidney stone, diabetes, cure of piles and bleeding caused due to piles,

					urine retention, rheumatoid arthritis, joint pain, purification blood
38	*Senna auriculata*	Avarike	Fabaceae	Whole plant	The leaves have laxative property, conjunctivitis, rheumatism, eye diseases,gonorrhoea and gout
39	*Solanum indicum* L	Kirigulla	Solanaceae	Root, fruit	Fever, dropsy, skin diseases, piles, eye diseases, chest pain, hard urination, sore throat, asthma, cardiac disorders
40	*Solanum nigrum* L	Ganike	Solanaceae	Leaves, roots, berries, stem	Bronchitis, itching, asthma
41	*Solanum xanthocarpum* Schard & H. Wendl	Nelagulla	Solanaceae	Fruit	Bronchitis, asthma, chronic
42	*Sauropus androgynous* (L) Merr	Chakramuni	Phyllanthaceae	Leaves, stem	Anti-diabetes, cough, anti-inflammatory, lungs problem, fever, eye infections
43	*Syzygium cumini* L	Jampunerale	Myrtaceae	Fruit	Diabetes
44	*Tagetes erecta* L	Chendhuvu	Asteraceae	Leaf	Ear seepage
45	*Terminalia arjuna* Q & A	Matti	Combretaceae	Bark	Skin diseases, wounds, tumours, peptic ulcers, wound skin ulcers, chronic low-grade fever, bone fractures, angina pectoris, myocardial infarction, coronary artery disease, heart failure, atherosclerosis
46	*Terminalia chebula* Retz	Alalekayi	Combretaceae	Leaves, seeds	Fever, loss of appetite, cough, asthma, obesity,jaundice, digestion problem
47	*Tinospora cardifolia* (Thunb)	Amrithaballi	Menisperm-aceae	Leaves, stem	Diabetes, high cholesterol, allergy, rhinitis, stomach pain, lymphoma, cancer, rheumatoid, arthritis, hepatitis, peptic ulcer, gonorrhoea, fever, bone fracture
48	*Tylophora Indica* (Burm.F) Merill	Adumuttada-balli	Asclepiadaceae	Leaves, root	Asthma, bronchitis
49	*Vitex Negundo* L	Lakki	Verbenaceae	Whole plant	Anti-cancer, rheumatism, cough, herbal medicine for women menstrual cycle

50	*Withania somnifera* (L) Dunal	Ashwagandha	Solanaceae	Leaves, roots	Inflammation, bronchitis, asthma, ulcer, hypertension
51	*Zingiber Officinale*	Shunti	Zingiberaceae	Rhizome	Fever, bronchitis, asthma, cough, digestion problem, diarrhoea, piles, stomach ache, cardiac diseases, anaemia

5. Eastern Zone

This zone encompasses tribes of **Jharkhand, West Bengal, Odisha and Bihar**. Major tribes of Odisha are **Juang, Kharia, Khond and Bhumij**; Jharkhand are **Munda, Oraon, Santhal, Ho and Birhor** which belong to Austric language family and speak **Kol and Munda** languages (Fig. 6.7). Ethnobotanical aspects of major tribes of eastern zone are given below:

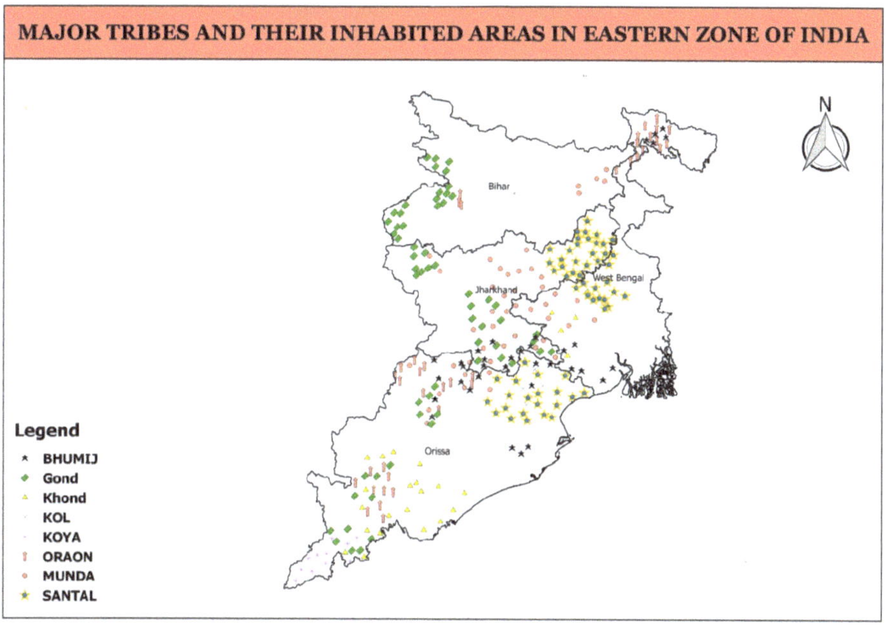

Fig. 6.7

Table 6.5 : Priority species used for traditional healing practices in Eastern zone of India.

S. No.	Botanical name	Local name	Family	Parts used	Ailments
1	*Abroma augusta* Linn.	*Ulatkambal*	*Sterculiaceae*	Root	Blood dysentery/ Diarrhoea/ Night wetting
2	*Acacia concinna* DC.	*Sikaki*	*Mimosaceae*	Leaf, seed	Dandruff / Premature hair greying
3	*Acacia leucophloea* (Roxb.) Willd.	*Arimed*	*Mimosaceae*	Bark or latex	Food poison
4	*Acorus calamus* Linn.	*Boch*	*Araceae*	Rhizome	Louse infestation
5	*Aerva lanata* Juss. ex Schult.	*Dayaluflower*	*Amarantha-ceae*	Flower & leaf	Asthma
6	*Alangium salviifolium* (Linn. f.) Wang.	*Ankar*	*Alangiaceae*	Root	Hydrophobia
	Aloe vera Linn.	*Ghritakumari*	*Liliaceae*	Fresh leaf	Burn, Rough skin
8	*Argemone mexicana* Linn.	*Shialkanta*	*Papavera-ceae*	Seed, plant latex	Impotence
9	*Aristolochia indica* Linn.	*Iswarnul*	*Aristolochiac-eae*	Bark, root	Snake bite, Food poisoning in cows
10	*Artemisia vulgaris* Linn.	*Nagdona*	*Asteraceae*	Leaf, root	Anti-fertility
11	*Artocarpus heterophyllus* Lam.	*Kanthal*	*Moraceae*	Peduncle	Snake's bite (Boas)
12	*Bambusa vulgaris* Schrad.	*Bans*	Poaceae	Root	Hydrophobia
13	*Brassica campestris* Linn.	White *Sarisha*	Brassicaceae	Seed	Acne / Alopecia
14	*Caesalpinia crista* Linn.	*Karanj*	*Caesalpinia-ceae*	Seed oil, Leaf	Alopecia / Boils/ Wounds Diabetes mellitus
15	*Cinnamomum zeylanicum* Breyn.	*Daruchini*	*Lauraceae*	Bark, roots	Mis-carriage (abortion)
16	*Clitoria ternatea* Linn.	*Swet aparajita*	*Fabaceae*	Root; juice	Infertility
17	*Cucumis sativus* Linn.	*Sasha*	*Cucurbitaceae*	Fruit; juice	Tumour
18	*Cuscuta reflexa* Roxb.	*Swarnalata*	*Cuscutaceae*	Stem; juice	Diarrhoea in cows
19	*Datura metel* Linn.	*Datura*	*Solanaceae*	Leaf	Alopecia
20	*Desmodium gyrans* DC.	*Bon-charal*	*Fabaceae*	Leaf	Enlargement of liver/ weak eyesight
21	*Eupatorium ayapana* Vent.	*Ayapan*	*Asteraceae*	Leaf	Acne
22	*Flacourtia indica* Merr. (Burm. f.)	*Baichi*	*Flacourtiaceae*	Stem, bark	Wasp bite
23	*Garcinia mangostana* Linn.	*Mangustan*	*Clusiaceae*	Leaf, husk; juice	Accumulation of fat
24	*Gymnema sylvestre* R.Br.	*Gudmar, Meshsringi*	*Asclepiadaceae*	Leaf, fruit	Diabetes mellitus
25	*Hemidesmus indicus* R.Br.	*Anantamul*	*Asclepiadaceae*	Root	Miliaria rubra (Miliary)
26	*Ipomoea paniculata* R.Br.	*Bon-kalmi*	*Convolvulaceae*	Leaf	Low blood pressure
27	*Jatropha curcas* Linn.	*Bherenda*	*Euphorbiaceae*	Fresh latex	Cuts and wounds
28	*Lycopersicon esculentum* Mill.	*Bilati Begun*	*Solanaceae*	Fruit	Tumour

29	*Malachra capitata* Linn.	*Bon-dhenros*	*Malvaceae*	Fresh fruit	Infertility
30	*Mimosa pudica* Linn.	*Swet-lajjabati*	*Mimosaceae*	Root; juice	Infertility
31	*Murraya koenigii* (Linn.) Spreng.	*Currypata*	*Rutaceae*	Fresh leaves	Diabetes mellitus
32	*Murraya paniculata* (Linn.) Jack	*Kamini*	*Rutaceae*	Leaf; powder	Cuts and wounds
33	*Nerium indicum* Mill.	*Shet Karabi*	*Apocynaceae*	Root; juice	Epilepsy
34	*Phyllanthus niruri* Hook. f. n	*Bhui Amla*	*Euphorbiaceae*	Whole plant; juice	High Bilirubin (Billiousness)
35	*Piper betle* Linn.	*Pan*	*Piperaceae*	Green leaves	Weak nerves of leg
36	*Piper nigrum* Linn.	*Golmorich*	*Piperaceae*	Fruit; powder	Constipation

6. Western Zone

The tribes of **Chhattisgarh, Madhya Pradesh, Western Rajasthan, and northern Andhra Pradesh states come under this zone.** The foremost tribes of Rajasthan are **Bhil, Garasia, Mina, Banjara, Sansi and Saharia**; of Gujarat are **Mahadeokoli, Bali and Dabala**; of **Madhya Pradesh** are **Jayanti (Fig. 6.8)**. Ethnobotanical practices of majour tribes of western zone are given below :

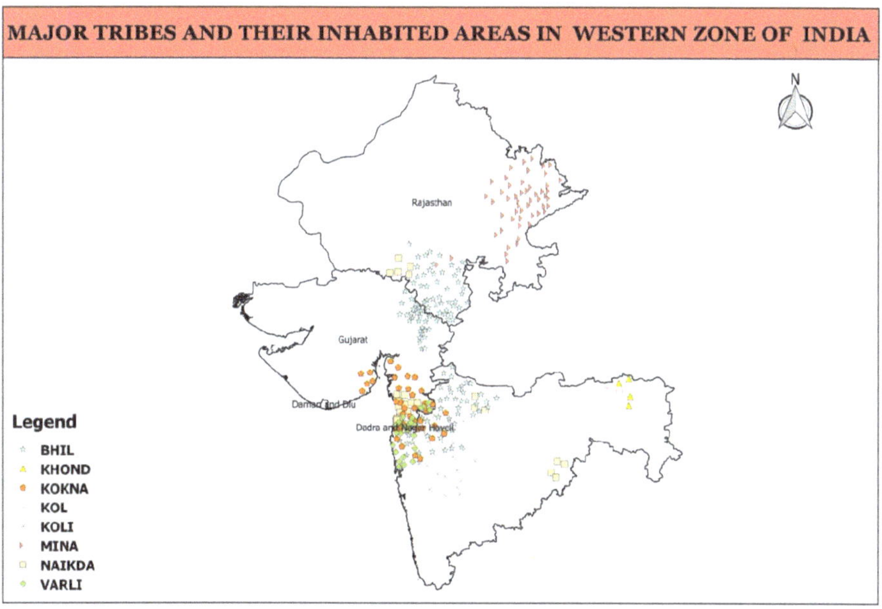

Fig. 6.8

Table 6.6.: Priority species used for traditional healing practices in Western Zone of India.

S. No.	Plants Name	Local name	Family	Part used	Medicinal Uses
1.	*Abutilon indicum* Sweet.	Tarakanchi	Malvaceae	Leaves, roots, bark	Uterine haemorrhagic discharges febrifuge, anthelmintic, diuretic, alexiteric.
2.	*Acacia nilotica* Linn	Babool	Mimosaceae	Bark, gum	Astringent, demulcent, asthma, diarrhoea, dysentery, diabetes.
3.	*Acacia Senegal* willd	Kumbat	Mimosaceae	Gum	Demulcent, emollient, haemorrhage, intestinal mucous, food for diabetes.
4.	*Acalypha indica* Linn	Khokali	Euphorbiaceae	Whole plant	Diuretic, bronchitis asthma, pneumonia, anthelmintic, rheumatism,
5.	*Achyranthus aspera* Linn	Andhi- jaro	Amaranthaceae	Whole plant	Laxative, skin diseases, eruption, colic, gonor rhoea
6.	*Adhatoda zeylanica* Need	Adusa	Acanthaceae	Whole plant	Emetic, dysentery, blood circulation jaundice, leprosy heart troubles, loss of memory
7.	*Aerva lanata* Juss	Gorkha-bundi	Amarathaceae	Whole plant	Diuretic, demulcent, diuretic.
8.	*Ailanthus excelsa* Roxb	Arru	Simaroubaceae	Bark	Appetizer, anthelmintic febri fuge, antispasmodic, diarrhoea, dysentry, chronic bronchitis recta
9.	*Balanities aegyptiaca* Del	Hingoto	Balanitaceae	Roots, Brak, seed	Anthelmintic, alexipharmic, antidysentric, analgesic, leucoderma, ulcers, skin diseases. purgative, anthelmintic. colic
10.	*Barleria cristata* wight	Raktajhinti	Acanthaceae	Whole plant	Inflammation, fever, bronchitis biliousness, tympanitis
11.	*Boerhavia diffusa* Linn	Punarnava	Nyctaginaceae	Whole plant	Astringent, biliousness, anaemia, leucorrhoea, inflammation, blood purifier
12.	*Boswellia serrata* Roxb	Salar	Burseraceae	Bark, Flower, Gum	Biliousness, dysentry, skin diseases, ulcer, blood purifier. leucoderma, pile

					santipyretic, skin & blood diseases.
13.	*Buchanania linsang* Spring.	Chiraunji	Anacardiaceous	Whole plant	Biliousness, blood diseases. expectorant, blood purifier. laxative, aphrodisiac, ulcers. cardiac tonic
14.	*Canscora decussate* Roxb.	Kalameg	Gentianaceae	Whole plant	Epilepsy, nervine debility, nerve tonic
15.	*Capparis deciduas* Edgew.	Ker	Capparaceae	Whole plant	Carminative, aphrodisiac, appetisor, emmenagogue, alexipharmic, lumbago rheumatism, hiccup
16.	*Capparis spinosa* Linn	Kabar	Capparaceae	Root bark	Emmenagogue, anthlmintic, diuretic, expectorant, analge sic, rheumatism, paralysis, toothache
17.	*Cassia fistula* Linn.	Amaltas	Caesalpiniaceae	Whole plant	Syphilis, leprosy, skin diseaseslaxative, antiperiodic, rheumatis manti pyretic, leprosypurgative, antipyretic, leprosy,heart diseases.carminative, appetizer
18.	*Catharanthus roseus* (L).Don.	Sadafuli	Apocynaceae	Leaf	Muscular pains
19.	*Cissampelos pareira* Linn	Akauadi	Menisperma-ceae	Root	diuretic, emmenagogue, dyspepsia, diarrhoea, dropsy, cough
20.	*Datura metel.* L.	Kala-Dhotra	Solanaceae	Seed	Cure asthma
21.	*Diospyros melanoxylon* Roxb	Timru	Ebenaceae	Whole plant	Diarrhoea, dyspepsia, ophthalmia, trichiasis. aphrodisiac, leucorrhoea, night blindness, spleen enlargement. digestive, carminative,
22.	*Ehretia aspera* willd.	Tombolan	Ehretiaceae	Flower	Venereal diseases.
23.	*Eruca sativa* Mill	Taramira	Brassicaceae	Leaves, seeds	Stimulant, stomachic, diuretic, antiscarbutic. aphrodisiac
24.	*Euphorbia hirta* Linn	Dudhali	Euphorbiaceae	Leaves	dysentry, diarrhoea, leucorrhoea, menorrhagia
25.	*Feronia elephantum* Curr.		Kaitha	Rutaceae	Leaves, seed and flower

					Hiccough, dysentry.alexiteric, heart diseases.aphrodisiac, alexipharmic, heartcomplaints, dysentry, leucorrhoea,opthalmia, billiousness
26.	*Flacourtia indica* Merrill.	Kanju	Flacoutiaceae	Flower	Appetizer, jaundice, spleen enlargement
27.	*Gloriosa superba* Linn	Kaligari	Liliaceae	Roots, tuber	Bowel complaints. anthelmintic, laxative, alexiteric, abortificient, ulcer, leprosy, piles
28.	*Indigofera tinctoria* wight	Neel	Papilionaceae	Roots, Whole plant	Laxative, expectorant, alexipharmic, anthelmintic leucoderma, abdominal complaints, heart disease bronchitis, asthma, piles,
29	*Kochia indica* wight	Bui	Chenopodia-ceae	Whole plant	Cardiac stimulant.
30.	*Melia azadarach* Linn	Bakain	Meliaceae	Roots, leaves & Stem	Anthelmintic, biliousness, tumors heart pain, leucoderma, ulcer, lung complaints, blood purifier. spleen & heart complaints dental problems rheumatism
31.	*Moringa oleifera* Lam	Sanjna	Moringaceae	Flower, fruits	Aphordisiac, alexiteric, analgesic, anthelminitic ulcers heart troubles,. ophthalmia, hallucination muscular & spleen siseases
32.	*Oroxylum indicum* Bent	Adhvantasha-trava	Bignoniace	Root bark, flower	Astringent, aphrodisiac, tonic, biliousness, bronchitis, worms. anthelmintic, stomachic, piles leucoderma, heart diseases.
33.	*Oxalis corniculata* Linn	Kharto	Oxalidaceae	Whole plant	Appetizer, dysentry, diarrhoea, piles, skin diseases.

34.	*Phyllanthus emblica* Linn	Aonla	Euphorbia-ceae	Fruit, leaves	Carminative, cooling alternative, alexiteric ulcer, bladder stone, bronchitis ophthalmia aphrodisiac, antipyretic, asthma, biliousness, leucorrhoea
35.	*Physalis minima* L.	Popati	Solanaceae	Fruit	Liver disorders
36.	*Plumbago zeylanica* Linn	Chitrak	Pumbaginaceae	Root bark	Dysentry, leucoderma, piles, inflammation, rheumatism bronchitis, anaemia liver & intestinal complaints
37.	*Psoralea corylifolia* Linn	Bavchi	Papilionaceae	Root, leaves, flower	Dental diseases diarrhoea diuretic, leprosy, skin diseases,asthma, bronchitis, piles.
38.	*Salvadora persica* Dence	Kharo-jal	Salvadoraceae	Leaves, flower	Astringent, anthelmintic, diuretic, analgesic, liver tonic.
39.	*Sida acuta* Burm.	Bal	Malvaceae	Roots, leaves	Aphrodisiac, demulcent in gonorrhoea, astringent, nervous & urinary trouble anthelmintic
40.	*Solanum nigrum* Linn	Makoi	Solanaceae	Flower, seed	Laxative, giddiness, gonorrhoea
41.	*Tecoma stans* (L). Juss. ex. Kunth	Sukali	Bignoniaceae	Root	Relieve pains
42.	*Tecomella undulate* Seem	Rugtrora	Bignoniaceae	Flower, seed, bark	Abdominal & liver complaints, laxative, anthelmintic, ulcers, blood & eye diseases.
43.	*Vernonia cinerea* Less.	Sandri	Asteraceae	Whole plant	Diaphoretic, strangury, promotes perspiration, spasm of bladder.conjuntivitis
44.	*Withania somnifera* Dunal.	Asgandh	Solanaceae	Roots, tuber, seed	Alternative, aphrodisiac, narcotic, tonic.
45.	*Zizyphus nummularia* Burm.f	Jari-bor	Rhamnaceae	Leaves, fruits	Cough cold skin diseases, astringent, cooling. biliousness,sores, ulcerated gums

7. Island Region

The tribes of **Andaman & Nicobar and Lakshadweep** groups come under this zone. The major tribes of Andaman & Nicobar are **Shompen, Onge, Jarwa, Sentinali and Nicobarese** and they belong to Negrito racial group (Fig. 6.9). The detail ethnobotanical information practiced by island tribes of India are given below :

The Central Agricultural Research Institute, Botanical Survey of India and Society of A & N Ecology (SANE) have mounted efforts towards compilation, collection and collation of information about medicinal plants available in these islands. From Andaman and Nicobar Islands, a total of 536 species of ethno-medicinal plants belonging to 382 genera and 138 families of Angiosperms, Gymnosperms, Pteridophytes and Algae have been recorded. Some of the comman medicinal plants commonly used by the locals and tribes and those found in the mangrove areas are listed below:

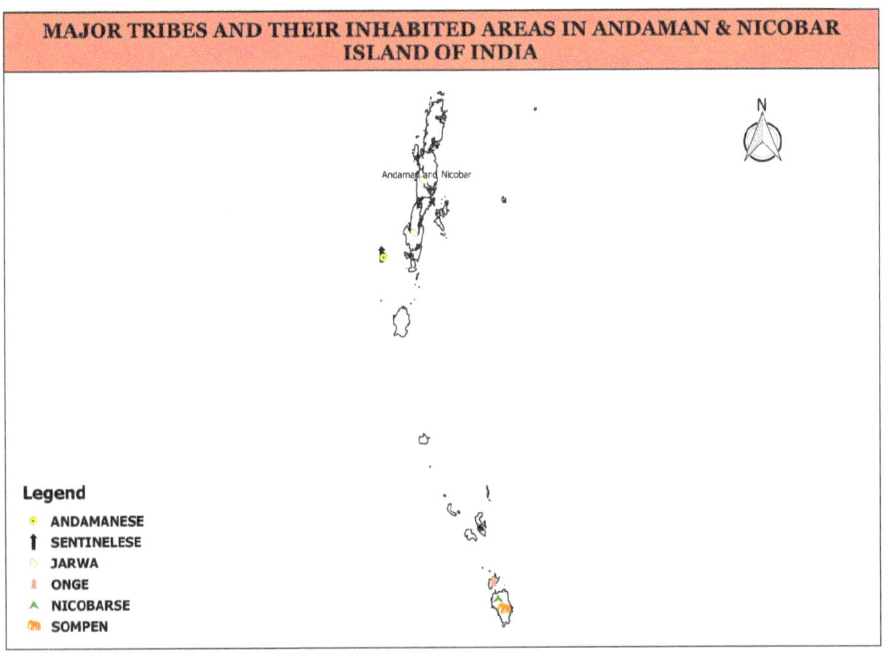

Fig. 6.9

Table 6.7 : Some Folk medicinal plants use by Negrito tribes of Island Region of India. (Source: Sharief 2007, Das *et al.* 2006; Dagar 1989; Chakraborty & Vasudeva 1988; Rao 1986)

S. No.	Plants Name	Local name	Family	Partused	Medicinal Uses
1.	*Alstonia macrophylla* Wall. Ex G. Don	*Chatium, Chuharoi*	Apocynaceae	Leaves, stem bark	Gastric disorder, malarial fever
2.	*Amomum fenzlii* Kurz.	*Jungli adhrak*	Zingiberaceae	Roots and flowers	Malarial fever, and gastrointestinal disorders.
3.	*Artabotrys nicobarianus D. Das*	-	Annonaceae	Leaves, bark,seeds.	Stomach pain, fever
4.	*Codiocarpus andamanicus (Kurz)R.A.Howard*	Kamarang	Icacinaceae	Leaves	Renovating agent, applied on enlarged scrotum of children
5.	*Costus speciosus* (Cone.) J. E. Smith	*Keokanda*	Costaceae	Rhizomes	Stomach disorders, headache and snake bite
6.	*Cytandromoea nicobarica Balakr.*		Scrophulariaceae	Leaves	Jungle fever
7.	*Dioscorea vexans* Prain & Burk	*Getti*	Dioscoreaceae	Tubers	Arthritis, asthma, eczema, chronic cough, iarrhea and diabetes
8.	*Donax canaeformis* (G. Forst.) K. Schum.	*Kagle, Tambowka*	Marantaceae	Root, stem, leaf	Abdominal & spinal pain, fever
9.	*Drypetes andamanica (Kurz) Pax & K.Hoffm.*	Toirulelu	Euphorbiaceae	Leaves	Chest pain
10.	*Etlingera fenzlii (Kurz) Škornick. & M.Sabu*	Jungli adhrak Hami	Zingiberaceae	Rhizome	Stomach disorders, fever;cough, fever, respiratory disorders and skin diseases
11.	*Hernandia ovigera* Linn.	Jack in the box plant	Hernandiaceae	Leaves, seeds, roots, bark	Headache and in treating tumour, epilepsy, fits
12.	*Hippocratea andamanica King*		Hippocrateaceae	Root	Ringworm, post-natal, rheumatism
13.	*Horsefieldia glabra* (Bl.) Warb	*Jugane*	Myristicaceae	Fruits	Abdominal pain
14.	*Kaempferia siphonantha* King ex Baker	Cekur	Zingiberaceae	Rhizome	Fever, stomach pain
15.	Kaempferia siphonantha King ex. Baker	*Cekur*	Zingiberaceae	Tuber	Fever and stomach pain
16.	Mallotus phillipensis (Lam.) Muell.-Arg	*Monkey face tree*	Euphorbiaceae	Fruits, seeds	Anthelmintic & abdominal pain
17.	*Myristica andamanica Hook.f.*	Kinhanmo	Myristicaeae	Fruits and leaves, seed, bark	Fever, malaria, stomach trouble
18.	*Phyllanthus andamanicus* Balakr. & Nair	*Dadaura*	Euphorbiaceae	Leaves	Antidiuretic
19.	*Premna serratifolia* L.	Tamonja	Verbenaceae	Fruits	Body pain, cough
20.	*Psychotria platyneura Kurz*		Rubiaceae	Leaves	General health problem
21.	*Scaevola sericea* Vahl.	Kwyae	Goodeniaceae	Fruits, Leaves	Rheumatic pain, fever, bodyache

22.	*Semecarpus kurzii* Engler	*Bara Bhilawa*	Anacardeaceae	Leaf juice	Anthelmintic, malarial fever, skin infection
23.	*Strobilanthesanda-manensis Bor*	-	Acanthaceae	Leaves	Blood in urine, asthma, cough, bronchial complaintst
24.	*Strychnos andamanensis Hill*	Lansot	Loganiaceae	Leaves	Urinary troubles, constipation, stomach ache
25.	*Tarenna weberaefolia (Kurz) Balakr.*	-	Rubiaceae	Leaves	Skin disease
26.	*Thottea tomentosa* (Bl.) Ding Hou.	*Udupet*	Aristolochiaceae	Whole plant	Chest pain, headache, cough, cold
27.	*Uvaria andamanica King*	-	Annonaceae	Leaves, flowers	Intestinal disorder
28.	*Wedelia biflora* DC.	Tukunkala, Tuvup-pukala	Asteraceae	Leaf paste applied	Cuts, headache

Some major tribes of India, their distributions, dilects and livelihood practices

India is ethno-linguistic composition of many tribes which are found across the various parts of the country (Fig. 6.10). The details of some of the major ethnic tribes mentioned in Schedule 5' of the constitution are given below:

Man of Chakma tribe of Tripura making bamboo crafts

Women of Santhal tribe of Odisha removing karnel of forest seeds

Woman of Mishing tribe of Assam collecting fish

Santhal tribe women of Odisha making pattals from leaves of sal

Woman of Chakma tribe tripura weaving cloth

Nyishi tribe woman selling forest produce

Woman of Mizo tribe Mizoram
selling forest produce

Nepali community in Manipur

Ao Tribe woman of Nagaland
collecting fuel wood

Children of Garo tribe of Meghalaya

Bhils (means Bow)

They are the third largest tribe in India after Gonds and Santhals, found in Rajasthan, Gujarat, MP, Maharashtra, AP and North-Eastern part of Tripura. Bhil tribe is divided into two main groups: Central or pure Bhils and Eastern or Rajput Bhils. Earlier they were great hunters, but now practices agriculture as the souce of livelihood.They speak Bhili, which is an Indo Aryan language.

Munda (means headman of the village)

They are the largest tribes of India, found mainly in Jharkhand but are also Spread across Bihar, WB, Chhattisgarh and Orissa. The Munda tribe is divided into following sub groups Ptat, Mahali and Kampat Munda etc. They remained hunters for centuries, but now they have been converted into the settled agriculturist. They speak Mundari, Santali, Munda or Killi languages, which is an Austro-Asiatic language.

Santhals

It is one of the largest tribe of India, found in West Bengal, Bihar, Orissa Jharkhand and Assam.They are mainly forest dewllwers and fulfilled their basic needs from trees and plants of the forests. In addtition they are also engaged in hunting, fishing and cultivation for their livelihood. They speak Santhali, which is an Austro- Asiatic language.

Gonds (mena hill people)

They were widely spread in the Chhindwara district of Madhya Pradesh, Bastar district of Chhattisgarh and also in the parts of Maharashtra, Andhra Pradesh, and Orissa.They are divided into four tribes namely-Raj Gonds, Madia Gonds, Dhurve Gonds, Khatulwar Gonds. They are mainly farmers and keeps herding cattles as part of their livelihood. They grow tobacoo for smoking and make liquor form mahua trees. They speak Gondi language which is related to the Telgu and the other Dravidian languages. In the northern parts Gonds are often seen speaking the local Hindi.

Khasi

Khasi were originated from Mayanmar and settled down in the plains of Assam. Mainly found in the state of Assam and the Jaintia hills in Meghalaya, other places are Punjab, UP, Manipur, WB and J&K. They are called by the different names such as Khasi Pahris, Khuchia, Kassi, Khashi and Khasa. There main occupation is Jhum cultivation (Shifting cultivation) and their staple food is rice. They speak Khasi-an Austro-Asiatic language.

Bodo

They are ethnic tribal communities of northwestern parts of Assam and Kachari is the sub branch of the Bodo. They mainly resides in the Brahmaputra valley and cultivate rice, pre Zu Mai (wine prepared from rice) and also rear silkworm. Their main dilects is Tibeto which is a Burmese language.

Gujjar/ Gurjara

They are mostly residing in the north western part of Himachal Pradesh, J&K, Punjab, UP, Hrayana, Northern Madhya Pradesh, Uttarakhand, Rajasthan, Gujarat and Maharashtra. Their occupation is semi nomadic herd animals cush as sheep, goats and buffalo. Baiga cuisine primarily consists of coarse grains, such as kodo millet and kutki, and involves very little flour.They speak Gojri, Punjabi, Hindi, Urdu etc. language.

Baiga

They are an ethnic group confined to central India primarily in the state of Madhya Pradesh, Jharkhand, Madhya Pradesh and Orissa.They are mostly forest-dwelling and do not interact with other tribal. They believe in a hand-to-mouth existence and practicing shifting cultivation. Their mother tongue is Naigani which as is an Austroasiaatic language.

Birhor (it means peoples of jungles)

They are confined to Chhattisgarh, Jharkhand, and Orissa and are known as the hunter tribe. This tribe is a Proto-Australoid tribe. The tribe is divided into two major sub-tribes into: the Jaghis Birhor are wandering in nature and practice shift cultivation while Uthalu Birho is a settled tribe and follow agro-based economy. They speak the Birhor language, which belongs to the Munda group of languages of the Austroasiatic language family.

Chenchus

They are native to central hill stations of Andra Pradesh, inhabit Nallamalai hills of Andhra Pradesh, Karnataka and Odisha. They are referred to as one of the Primitive Tribal Groups that are still dependent on forests and do not cultivate land but hunt for a living. They speak Chenchu language with Telgu accent.

Gaddis

They mainly dwell around Himachal Pradesh in the regions of Dhauladhar mountain range, Chamba, Bharmaur and the areas near to Dharamshala. Their main occupation is shepherding and they make their livelihood by rearing and selling sheep, goats, mules and horses. They speak the Gaddi Language but for writing they use Takri and Hindi.

Pangwal

They are the inhabitants of the Pangi valley of Chamba district in Himachal Pradesh. They are mostly engaged in farming.

Bhutia

They are of the Tibetan origin and migrated to Sikkim around 16th century. They are also known as the Lachenpas and Lachungpas. Majority of them concentrated in the dry valley of North Sikkim, hill districts of West Bengal such as Darjeeling and Kalimpong. They form 14% of the total population of Sikkim and eat rice, meat and beef. Their main language is Sikkimese but also speak Lepcha, Bhutia, Nepali, English and Hindi.

Limbus / Limboos

They are one of the indigenous tribe of Sikkim belonging to Nepali stock. Mongoloid looking by figure having their own language, faith, costume, culture and life style.They are farmers by profession.

Lepchas (means Children of God)

They are one of the indigenous tribe of Sikkim. They call themselves Rongkup and their language is Lepcha. They follow the same Tibetan form of Buddhism besides their local faith.

Buksa / Bhoksa (means Phari goat)

They are one of the primitive tribal groups of India found in Uttarakhand and UP, mainly concentrated in Dehradun and Nainital districts in Kumaon region and Bijnor disrcts of Uttar Pradesh. Their major occupations are agriculture and animal husbandry. Tharu speak Thry language which is an Indo-Aryan language.

Apatani or Tanni

They are settled agriculturists inhabiting the valley around Ziro-the headquarters of Lower Subansiri district in Arunachal Pradesh. They practice permanent wetland cultivation in hilly terrain by slicing the hills.

Adis / Bokar Lhoba

This tribe is found in the region of Arunachal Pradesh, Assam and is divided into two main such as Bogum and Bomis.They by nature are democratic and organized village council called Kebang.

Nyishis

In Arunachal Pradesh they constitute a large group of people inhabiting the major part of Lower Subhansiri district.They are Indo-Mongoloid group of people and their language belongs to the Tibeto-Burman family.The tribe as a whole is fond of hunting and fishing in which they perhaps excel than any other neighboring tribes.

Kampti

They are a sub-group of the Shan people in Lohit district in Arunachal Pradesh. The Tai-Khamti have their own script for their language, known as 'Lik Tai', which was originated from the Shan (Tai) script of Myanmar.

Tagin

They are main inhabitant of Upper Sunansiri district in Arunachal Pradesh. Their main occupation is agriculture.

Asur

Usually found in the sate of Jharkhand in the eastern part of the subcontinent of India. There main occupation is agriculture and eats rice, flesh of birds and animals. Their mother tonguge is Hindi.

Angami

The Angamis are a major Naga ethnic group native to the state of Nagaland in North-East India.Tenydie is the most common language spoken. Agriculture is their main occupation.

Rengmas

They are one of the seventeen major Naga Tribes located in Nagaland and section of them have migrated to Karbi Anglong districts of Assam. They follow patriarchal system. Originally they were animist and Agriculture is the main occupation.

Sema / Sumi Nagas

They are one of the major tribes of Nagaland recognized for their martial art skills. They mainly inhibit Zunheboto in Nagaland and Assam. They are most united and aggressive tribe. Their diet consist of mainly rice, millet accompanied by vegetables, fish and meat. They speak 60 different dialects belonging to the Sino-tibetan family of languages.

Zeliang

They live primarily in the south-western part of Kohima district and also inhabit Barrail range of hills in Manipur, Nagaland and Assam.They speak Zeme which is a Sino-Tibetan language.

Konyak (means blackhead or human)

They are the largest out of 17 officially recognized tribes in Nagaland. In Nagaland, they inhabit the Mon District also known as 'The Land of The Anghs'. They are known as 'those violent head hunters with tattooed faces.' They now practice agriculture and hunt seasonally. They speak Konyak language which belongs to the Northern Naga sub branch of the Sal subfamily of Sino-Tibetan.

Garos

The Garos are an aboriginal Tibeto-Burman ethnic group from Indian subcontinent, found in Meghalaya, Assam, Tripura, Nagaland. They are the second largest tribe in Meghalaya after the Khasi and comprise about a third of the local population. Their language, Khasi, is categorised as the northern most Austroasiatic language.

Jaintias

They are traced in the Jaintia hills in Meghalaya, belong to the Proto Autroloid Monkhmer race. Basically it is a sub tribe of Khasis and since they have been residing in Jaintia hills, they have been named so.

Reang

They are the second largest of tribal community Tripura who came from Shan State of upper Burma and belong to Indo-Mongoloid racial stock. Reang tribe is nomadic in nature and a large numbers among them maintain their livelihood through Top Hill Jhum Cultivation, food gathering, fishing & hunting. Their language is known as "Kaubru" which have a tonal effect of Kuki language though broadly it is Kok-Borok (language of men).

Lushais

They belong to Kuki-Chin group of tribes in Tripura. They live on Jhum Cultivation and hunting of wild animal. They are also famous as orange producing community and speak Kuki-Chin languages.

Jarawa (means stranger)

The Jarwa are found in south and middle Andaman and live as nomads in a group of 40-50 people. They are hunters and gather and eat wild pigs, moster lizards, fish, fruits and berries. They mainly speak Andamanese language.

7

Ethnobotany of Plants from Major Families

This chapeter deals with indigenous knowledge on use of various plant species in traditional medicine, food, dye, wood and in tanning etc.

Malvaceae

The family Malvaceae is also known as Mallow family, characterised by mainly shrubs and herbs. The plants of this family frequent have stellate trichomes, persistant sepals, epicalyx present, stamens numerous and monadelphous, anthers monothecous, carpels 5 or more.

Abroma angusta (Devil's Cotton)

Paste of the leaf is applied to heal wounds and to relieve joint pain or back pain. Young shoots are used as food. Roots are used to cure itching.

Abutilon indicum (Kanghi)

The whole plant is used to make brooms. Plant also has medicinal value and its sap is used against ulcer, leprosy, piles, dysentery and fever.

Althaea rosea (Hollyhock)

The roots of this plant is used in dysentary and leaves is used to produce blue colour dye.

Hibiscus rosa-sinensis (Gudhal)

Planted as ornamental plant. Roots are used in fever, rheumatism, diabetes, eye complaints, cough and cold.

Malva verticillata (Chinese Mallow)

Leaves and seeds are edible. Seeds and leaves yield green and yellow dye. Various parts are used aginst whopping cough, renal disorders and diarrhoea.

Ebenaceae

The family Ebenaceae includes trees and shrubs which are mainly dioecious. The leaves are simple, alternate, inflorescence cymose, flower having 3 to 8 basely joined petals, several single or paired stamens often attached to the inner wall of the corolla and fruits are berry-like or capsular.

***Diospyros melanoxylon* (Tendu/Biripatta)**

The roots of tendu are dried under sun light and one teaspoon of decoction prepared from pounded bark is taken twice a day up to 5-7 days to gives relief from diarrhea. Fruits are edible.

***Diospyros malabarica* (Indian Persimmon/Gaab)**

Wood is used in construction work, furniture and carving. Leaves are used for making beedi. Branches are used as fire wood, fruits are eaten by tribals. All parts of the plant are very useful and used in treatment of various diseases like diarrhea, leprosy, gonorrhea etc.

***Diospyros montana* (Bombay ebony/ Bistendu)**

The wood is used to make agriculture implements, musical instruments, furniture and frames. Leaves are used as fish poison; fruits are poisonous and used to treat burns.

***Euclea natalensis* (Gaurri)**

Black color dye is obtained from root. Root is also used as tooth-brush. The decoction of root is used to kill parasitic worms, asthma, bronchitis, indigestion and diarrhoea.

Rubiaceae

The family Rubuiaceae is also known as Madder family, characterised by herbs, shrubs or trees ; leaves stipulate, opposite or whorled ; flower terta or pentamenous ; stamens alternipetalous and as many as corolla lobes ; usually bicarpellary; ovary inferior.

***Chinchona officinalis* (Quinine)**

It is a medicinal plant, famous througout the world for production of quinine from bark, which is an anti malarial agent. Other alkaloids that are extracted from bark of this tree include cinchonine, cinchonidine and quinidine.

***Ixora coccinia* (Lal Lokhandi)**

Leaves and flowers are used in treatment of dysenatry and diarrhoea.

***Gradenia gummifera* (Dekamali)**

It provides gum which is used as a carminative stimulant.

***Randia spinosa* (Madanphala)**

Various plant parts are used traditional in treatment of various disease like stem bark is used in dysentery and diarrhoea. Fruits are used as a remedy in ulcer, tumor and skin complaints and gastro-intestinal problems.

***Coffea arabica* (Coffee)**

The seeds of *Coffea arabica, C. liberica* and *C. robusta* are roasted and ground to give coffee powder.

***Morinda citrifolia* (Noni)**

The various parts of Noni tree like stem, leaf, bark and fruits are used in ailment of cough, cold, arthritis, liver disease, malaria, tuberclosis and blood pressure.

***Mussaenda frondosa* (Flag bush/Dhobi tree)**

Root and leaves are used in ulcer, cough, bronchitis, Jaundice and snakebite.

Guttiferae (recently Clusiaceae)

They are mostly trees and shrubs, with milky or colored sap, leaves simple and are opposite, whorled or rarely alternate, flowers actinomorphic and unisexual, the perianth consists of a calyx of 2-10 imbricated, often decussate sepals and 4-12 petals, stamens numerous and are distinct or variously united. The fruit is usually a capsule or berry.

***Garcinia cowa* (Brindal berry/Malabar tamarind)**

Leaves are used as fodder for cow and goats and cooked as vegetable also. Fruits are edible. Stem wood is used for light furniture. Branches are used as fire wood. Red or yellow dye is obtained from stem bark or fruit. Plant extract is used to treat indigestion, cough and cold.

***Garcinia dulcis* (Gaurka)**

The bark yields a black dye used for cotton.

***Garcinia pedunculata* (Tikul/Amlavetasa)**

Timber is durable and used for house construction, making furniture and agriculture tools. Fruits are edible and used treatment of various ailments in dysentery, diarrhea, weight loss.

***Mesua ferrea* (Ceylon Ironwood/ Nag kesar/ Nag champa)**

Leaves are used in cough and cold, seed oil to cure rheumatism and scabies. Flowers are used to treat indigestion and snake bite.

Calophyllum inophyllum (Alexandrian laurel/Sultan champa)

Bark yields dye, wood is water durable and used for construction of boats, flooring, cart wheels and furniture. Seed oil is used as insecticidal, antifungal, anti bacterial. Plant extract used on wounds, ulcer and in treatment of gonorrhea.

Fabaceae

Generally plants are herbs, shrubs and very few trees. Leaves are alternate or opposite, stipulate and petiolate. It composed five petals, corolla papilionaceous, aestivation vexillary, stamens 10, diadelphous or monoadelphous. Fruit is legume or pod and lomentum, splits along both sutures.

Abrus precatorius (Ratti/Gumchi)

Leaf and seeds are used in treatment of fever, tuberculosis, asthma, dog bite, cat bite.

Butea monosperma (Dhak/Palas/Tesu/Khakra)

Natural dye is obtained from the flowers, leaves are an excellent fodder or used for making *dona* (bowl), *pattal* (plates), stem yields gum that is known as *Kamarkas* and taken orally by the tribal or rural women after delivery. Fresh extract of flowers are used in gastric troubles.

Pterocarpus marsupium (Bijasal)

Wood is strong and durable, used for construction purpose, musical instruments, boat and railway carriages. Heartwood and gum yield tannin and yellow dye. Leaf, bark and plant sap is used in treatment of diarrhea, skin disease and diabetes.

Pueraria tuberose (Sural/Tirra)

Tuberous roots are used to cure cough, constipation, fever, rheumatism and as a diuretic.

Mimosaceae

Most of the plants are tree or shrub with bipinnate leaves. Stipules are modified into spines. Inflorescence is cymose head or head, petals connate below. Stemens varies from four to many and fruit type is legume or lomentum.

Mimosa pudica (Lajvanti/Touch me not)

Leaf and stem is used to cure hydrocele. Leaves after crushing used to stop bleeding form newly wounds.

Acacia modesta (Phulai)

Wood is used as fuel wood and in small construction works, leaves are used as fodder. Bark and various parts of the plant are used as a pain killer and dysentery.

Acacia catechu (Khair)

Wood is used in construction works, fuel wood, agriculture implements, Stem bark is used in treatment of Bronchitis, diarrhea, dysentery and cold. Katha obtained from heart wood used to cure mouth ulcer. Cutch is locally used for tanning leather.

Acacia senegal (Kumat)

Plant is used in fencing; wood is used as fuel wood. Gum obtained from kumat is used . Leaves are used as fodder and in various traditional medicines like in burns, bounds and inflammation.

Acacia concinna (Shikakai)

Young leaves, flower and pods are cooked as vegetables. Pods and seeds are used in treatment of dandruff and washing hairs and woolen clothes. Leaves are used as anti-malarial and bark in diarrhea and dysentery.

Prosopis cinearea (Khejri)

Leaves are used as fodder, pods as vegetable and wood in various construction works, agriculture implements. The stem bark is used as vermifuse and in treatment of asthma, leucoderma, dysentery, rheumatism and piles.

Caesalpinaceae

Mostly shrubs and trees, rarely herbs. Leaves are pinnately compound paripinnate, stipulate arranged alternately, Inflorescence raceme or panicle, stamens10 (5+5) with staminodes, fruit type is legume or lomentum.

Cassia auricultа (Tanner's cassia/Awal)

Whole plant is used in tanning, flowers and seeds are used to cure diabetes.

Cassia tora (Puad)

All parts of the plant are used in treatment of piles, dropsy, skin complaints.

Cassia occidentalis (Coffee Senna/Kasunda)

Young shoots are cooked as vegetable. Whole plant is used to cure skin disease, bone fracture, urinary complications, fever and skin disease.

Cassia fistula (Amaltas)

Flowers and fruit pulp is edible. Stem bark is used in various skin complaints Fruit pulp is used in treatment of dysentery, diabetes, anthrax and malaria. Roots are used in ulcer and fever.

Bauhinia vahlii (Mahul/ Camel's foot)

Stem bark is used in tanning, fibers are obtained from inner bark. Seeds are edible and used as aphrodisiac. Leaves are used in thatching and making plates and dona (bowl).

Bauhinia tomentosa (Pili kachnar/Yellow orchid tree)

Young leaves are cooked as vegetable. Purple dye is obtained from the flowers. Root bark is used as vermifuse and to treat various skin diseases.

Combretaceae

This is known as terminalia family, most of the plants are trees and shrubs, flowers pentamers, stamens 10 in two series fruit is single seeded drupe with 1-5 angled or winged pericarp. Tannin cells are present in whole plant body.

Terminalia chebula (Harad/Black myrobalan)

Timber is durable and used in construction of house and various implements. Seeds are consumed raw or fried. Fruits are used in tannin industry and ethnomedicinaly in stomachace, fever, cough and cold.

Terminalia arjuna (Arjun)

All parts are very useful in treatment of various disease like asthma, tumer, aenemia, dysentery, urinary problem and heart complains.

Terminalia bellirica (Baheda)

Wood is moderately hard and used for construction of agriculture implements. Bark is used for production of dye. Fruits and leaves are used for tanning purpose. Fruits are used for treatment of indigestion, sore throat, cough, diarrhea and an important gradient of triphala churna in ayurveda.

Terminalia tomentosa (Saja/Sadar)

Wood is used in house construction and as fuel wood in marriage ceremonies. The stem bark of Saja is used in liver complaints.

Umbelliferae (recently, Apiaceae)

Plants are annual herbs and perennial under shrub and shrubs. Leaves are pinnately compound alternate, exstipulate, amplexicaul and rarely opposite. Inflorescence is umbel or compound umbel, fruit type cremocarp.

Centella asiatica (Brahmi)

Leaves or whole plant is used to treat urinary complications, mental disorder, blood infection, rheumatism, cholera, piles and skin disease.

Coriandrum sativum (Coriander/Dhaniya)

The green leaves and dry fruits are used as condiment and spice. Dry fruits are also used as carminative, stomachache and tonic.

Daucus caryota (Carrot/Gajar)

Roots are edible raw or cooked. Leaves are used as fodder. Seeds are used as abortifacient, kidney stone, cough and bronchitis.

Bunium persicum (Kalajeera)

Fruits are used to treatment of indigestion, gastrointestinal pain, abdominal and colic pain.

Carum carvi (Jeera)

Seeds are used in ailment of indigestion, diarrhea, dyspepsia, inflammation and skin infections.

Asteraceae

Mostly plants are herbaceous and shrub rarely tree. Leaves are simple or scaly, alternate, opposite or whorled. Inflorescence is head or capitulum consisting disc or ray florets.

Artemisia maritima (Sea wormwood)

Locally used as insect repellant, to cure indigestion, headache, fever, neurological problems and as carminative, vermifuse.

Ageratum conyzoides (Jangli Pudina/Goat weed)

Stem and leaf extract used as antifungal, wound healing, fever, indigestion, leprosy and eye related disorders.

Eclipta alba (Bhriangraj)

It is well known for hair oil which is used for long and black hairs. Whole plant is used as diuretic, in treatment of jaundice, spleen enlargement and liver disorders.

***Gnaphalium hypoleucum* (Cotton weed/ Cud weed)**

Young shoots are cooked as vegetable. Whole plant is used in treatment of wounds, cough and backache.

***Sphaeranthus indicus* (Gorakhmundi)**

Whole plant is used in various traditional medicine including seeds and roots treat cough, fever, leprosy, diabetes, jaundice and mental illness.

***Tridax procumbens* (Talmuria/ Patharchatta)**

Whole plant after crushing used for stop bleeding, used to cure malaria and diarrhea. It is also used as a fodder for animals.

Apocynaceae

The family apocynaceae is also known as dog bane family, characterised by trees, shrubs, climber and herbs. The plants of dog bane family usually have milky often poisonous latax sap, smooth margined, opposite leaves, flowers in cluster.

***Alstonia scholaris* (Saptaparni/Devil tree/Chitvan)**

Stem bark is used in treatment of diarrhoea, malaria, fever, cholera, gastric complaints, headache and asthma. The bark is taken in post natal condition to increase lactation in mothers. The milky juice of stem is used in treatment of pimple and pyorrhoea. Decoction of leaves are used in beri-beri, ulcer and rheumatic pain. The flowers are used in treatment of asthma. White coloured light wood is used for making utensils and toys.

***Carissa carandas* (Karaunda/Christ's thorn)**

The wood is used for making kitechen items like spoon and the leaves are valued in fever, earache and diarrhoea. The fruits are used in curries, tarts and chutney. The green unripe fruits are used for making pickle. The unripe fruits are used medicinally as constringent. The fruits are alsoused as agents in tanning and dying. Roots are used as fly repellent.

***Catharanthus roseus* (Sadabahar/Vinca/Periwincle)**

The juice from the leaves is used to treat wasp stings. The whole plant is medicinally valued and used orully for cure of cancer, various skin problems, nose and gumbleeding, diabetesetc. The extract of root is recommended in menorrhagia.Vincais a diuretic and increase urine. Flower petals and seeds are used as an antioxidant.

Holarrhena antidysenterica (Karva Indrajao/Kutaj)

The stem bark is used to cure urinary troubles, diarrhea. The bark of *Holarrhena* helpful in checking blood and mucus in piles. The decoction of seed is recommanded in diabetes.

Ichnocarpus frutescens (Black creeper)

The leaves are boiled in oil and applied between fingers for the treatment of headache, fever and wounds. The seeds are used in ailment of rheumatism. The root is sweet in taste and used as demulcent, febrifuge, refrigent, aphrodisiac and diuratic. The complete plant is used for the treatment of night-blindness, bleeding gums, enlargement of spleen, ulcer, smallpox, asthma, snakebite and abdominal tumor.

Nerium oleander (Kaner/Oleander)

The leaves of oleander mixed with honey applied to treat scabies. Kaner is used by tribals as arrow and dart poisonsince long back ago. By considering of its toxic potential it has been used in treatment of cancer, asthm, aepilepsy and cardiacdisorders. In folk medicine the flowers are used in ailment of malaria, ringworm, leprosy indigestion and heart-falure.

Plumeria alba (Plumeria/Frangipani/Champa)

The stem bark is used in tounge cleaning and in treatment of stomachache, intestinal worms, rheumatism and malaria. The leaves juice is used in fractures, pods boiled in milk are used in snakebite. The latex is used in treatment of toothache, earache, eye cleaning liquid, sores and blisters caused by mosquito bite. The flowers are used in ailment of diabètes and dysentery. In Assam some ethnic communities used the flowers for permanent sterilization of female (Dey, 2015).

Rauvolfia serpentina (Sarpgandha)

The juice of leaves are used as an antidote against snakebite in some tribal rich areas of Madhya Pradesh, Orissa, Uttarakhand and Karnataka. The roots of the plant are used in high blood pressure, mental agitation, sédative and madness. hypertension, epilepsy and nervous disorders. The leaves after crushing used orally in dog bites.

Wrightia tinctoria (Indrajao, dudhi, khirni)

Blue colour dye is obtained from the seeds of Indrajao. Stem bark is used as paresthesias, intoxicant, antidysentric and leaves are used as laxative, anti-dandruff and treatment of toothache, fever and stomachache. Seeds are used in seminal weakness, diarrhoea and dysentry. The powder of its dry root with amla (*Emblica officinalis*) and samalu (*Vitex nigundo*) is taken with milk by women for increasing fertility (Srivastava, 2014).

Asclepiadaceae

Plants herbs, shrubs and mostly twinners, rarely tree. Leaves simple opposite and entire margin; stamens 5, fused with stigmatic disc forming gynostegium; pollens forming pollinia, fruit follicles.

Asclepias curassavica (Scarlet milkweed/Kakatundi)

Plant yields stem fibers, which are used to make coarse ropes. The soft seed hairs are used as stuffing material for pillow. The leaves after crush, mixed with common salt and vegetable oil used in treatment of skin ulcer. The latex is used on wart, sores, dermatitis and corns. The decoction of root is used to eye wash and dysentery.

Calotropis procera (Aak/Madar)

The powder of leaves is used in dysentery. The flower and roots after grinding and making powder are used in leucoderma and rheumatoid arthritis.The bark powder is used in treatment of elephantiasis. The latex is used in treatment of blisters, scorpion sting and eye disorders. The leaves of madar used as soup ingredient in various types of sauces.

Cryptolepis buchanani (Indian sarsaparilla/Kala bel)

This plant is used in treatment of cough, diarrhea, ulcer, inflammation, blood purifier, snake bite, dogbite and bacterial infections. The leaves are used to cure piles andbone fracture. The root and whole plant is used in treatment of syphilis and to increase lactation. Soft fibers are obtained from the stem that used in making cordages.

Cryptostegia grandiflora (Rubber vine/Rubber ki bel)

Latex is used for production of poor quality rubber and in poisonous biteof many insects. Root paste isused in treatment of asthma. Bark is used in influenza and whole plant is used as antibacterial and stimulant.

Oxystelma esculentum (Rosy milkweed vine/Dudhialata)

The latex is used as washing agent for ulcers. The plant is diuretic, aphrodisiac, laxative, bronchitis, leucoderma and in expelling of internal worms. The *Kada* prepared by the plant is useful in mouth and throat infection. The leaves are used in jaundice and root is very effective.

Gymnema sylvestris (Gurmar)

The tender leaves and fruits are cooked as vegetable or roasted. It is used to reduce the blood sugar. Both dried leaves and root are used orally for the treatment of diabetes, asthma, piles, jaundice, gastric troubles, water retention and obesity. Both stem and leaf sap is used as tonic.

Pergularia daemia (Utaran/Gadaria ki bel)

Milky latex is used in bone fracture, body pain, headache, stomachache, cough and cold, gastric ulcer, menstrual complications. The leaves are used after crushing in leprosy and amenorrhea and diarrhea. Stem bark is used as a appetizer, antipyretic and anti-malarial.

Sarcostemma acidum (Somlata)

The whole plant or latex is used to cure cough, fever and bronchitis, rheumatism, arthritis, earache, snake bite, mental disorders, lactation, rejuvenation, skin diseases, cataract, hydrophobia, scabies and ulcer.

Leptadenia reticulata (Jivanti)

It is an important medicinal plant of arid and semi arid regions. The leaves and root are used to treat various diseases like common fever, cough, ringworm, wounds, dysentery, night blindness, burning sensation and cancer. In the form of tonic it gives strength to the body.

Euphorbiaceae

Commonly called spurge family, mostly monoecious herbs, shrubs and trees characterize with milky sap. *Cyathium,* Special type of inflorescence in which a single naked pistillate flower is surrounded by cymes of naked staminate flowers.

Emblica officinalis (Amla/Aonla)

Brown colour dye is obtained from the fruits of Aonla. The leaves, fruits and bark is rich in tannin and used in tanning for animlal hides. Fruits are rich in vitamin C used in making pickle, murabba or dry candy etc. The fruits are used in treatment of dandruff problem, fever, ulcer, diarrhoea, inflammation, scurvy, nausea, skin disease and as antiaging and increase longevity. Bark is used in gonorrhoea and leaves are used as mouth wash, asthma, bronchitis and as antipyratic.

Ricinus communis (Castor/ Arandi)

The seed of castor are used as an anthelmintic, cathartic, emollient, laxative, purgative and in indigestion. The powder is applied to abscesses and various skin disorders. It is used as a purgative, discutient and expectorant. The leaves are used as a poultice to relieve headaches and treat boils.

Jatropha curcas (Jatropha/Ratanjot)

Traditionally, the seed oil of jatropha use in lightning the lamp as a substitute of kerosene oil. The thin branches are used as tooth brush for cleaning teeth. Decoction of leaves is used in fever, rheumatism, malaria, jaundice and mouth infection. It also used as pesticidal, coagulant, anticancer, purgative, gout, stomachache, venereal disease and as blood purifier. It has also been proved as an effective livestock barrier around the cultivation.

Jatropha gossypifolia (Jatropha/Jamalghota)

This plant is also planted as livestock barrier because of non preference of the animals for browsing. The leaves and seeds are used as purgative. Stem sap is used to cure common headache.

Mallotus phillippensis (Rohini/Kamala/Sinduri)

Orange colour dye is obtained from the outer surface of the fruits of Rohini tree. The stem bark yields fibers which is used for making cordage. Stem bark and leaves contain tannin which is used in tanning purposes. The leaves are used as an appetizer. Kamala is used as carminative, purgative, muturant and in enlargement of spleen.

Palmae (recently Arecaceae)

It is a family of monocotyledons that comprises woody shrubs, vines and trees, characterized by large or very large leaves with a tubular sheathing base, inflorescence is paniculate; fruit berry or drupe.

Areca catechu (Betel palm)

Betel palms are a good source of tannin and dye both. The leaves are used in making mats and thatches. Leaf sheath is used for making handicraft items like cups, donas, bags etc. The tender leaves, shoots and inflorescence are cooked as vegetable. The fruit (nut) of betel palm has multiple therapeutic properties like, astringent, diuretic, laxative and skin disorders. Root is effective against various ailments like urinary tract disorders, skin irritations, and as a component in health tonic preparation.

Cocus nucifera (Coconut/Nariyal)

The coconut fruits are eaten raw or in a variety of dishes. The liquid inside the fruitis taken as a cool drink. The tender inflorescence is used as vegetable. The coconut oil is used in reheumatism, back pain, diabetes, diarrhoea, cancer, etc. It is used to killing bacteria causing urinary tract infections, gonorrhea, gum disease and kidney inflammations. The coconut milk is used in treatment of diarrhea.

Phoenix dactylifera (Arabian date palm)

The leaves are used for making hand fans, baskets, bags and mats, thatches etc. The stem is used in construction of local houses. The dried fruits and tender inflorescence are consumed raw or after fry. The fruit of date palm is good in heart complaints, abdominal complaints, fevers, vomiting and loss of consciousness. The juice obtained from the tree is considered a cool and refreshing drink.The roots are used to treat toothache and gonorrhea.

Arenga wightii (Whight's sago palm/wild coconut)

The leaves of this tree are used in thatching and making mats. The tender leaves are cooked as a vegetable, starch from stem use to make *Upama* and palm wine is also obtained. Starch powder, obtained from stem pith is administered up to one week in leucorrhoea and painful urination (Udyan *et al.*, 2008).

Caryota urens (Fishtail palm/ Toddy palm)

Stem starch is edible used to make flour and bread. Paste of young plant stem along with the fruits of *Phyllanthus emblica* and rhizome of *Curculigo orchioides* is taken to strengthen the body (Ayannar and Ignacimuthu, 2005). Toddy (palm wine) is obtained from this tree.

Calamus rotang (Rotang/Rattan)

Tender shoots are used as vegetable. Stem is used in making furniture, walking sticks and other handicraft items. The plant is used in treatment of fever, piles, diarrhea, eye problems, killing intestinal worms and to increase saliva in chewing.

Lauraceae

Family lauraceae comprises woody flora; trees and shrubs except parasitic genus Cassytha. Leaves are simple, exstipulate, alternate, aromatic and entire. Essential oil is present in bark and leaves both. Flowers mostly solitary panicles, cumes or umbels, fruit drupe born on enlarge perianth looking like cup.

Litsea glutinosa (Maidalakri / Indian Laurel)

Wood contains gluten, used for joining the bone fracture. The seeds are used tribal and villagers for washing the clothes. The stem bark and leaves are used to cure fever, swelling, diarrhea, dysentery and to enhance sexual power. Coarse fibers are also obtained from roots for making cordage used for tying cattle.

Cinnamomum verum/zeylanicum (Dalchini)

The stem bark and leaves are used as spice and used in treatment of diabetes, cough and dysentery, nausea, cold, vomiting and flu. The essential oil present in stem bark and leaves used for flavouring various dishes. The cinnamon oil is applied externally to cure rheumatism, sore gums and toothache. The heartwood is used in construction and furniture.

Cinnamomum tamala (Tejpat)

Leaves are used as spice, essential oil of leaves used as diuretic, heart complications and in gastrointestinal complaints.

Cassytha filliformis (Love vine/Akas bel/dodder laurel)

It is only single semi-parasitic vine of family lauraceae which is (Wiry stem and tiny leaves) used in dysentery, liver infection, malarial fever, dandruff, gonorrhea, ulcer and eczema. Brown dye is also obtained from the stem.

Poaceae

Annual or perennial herbaceous plants rarely woody (Bamboo), stem cylindrical, nodes and internodes, leaves are sessile with sheathing base, parallel venation, inflorescence: spike of spikelets, flower pedicillate, two bract palea and lemma enclose flower, fruit caryopsis.

Cymbopogon martini (Palmarosa grass)

Traditionally the leaves of the plant are used in tea, insect and fly repellent. The essential oil is used in cosmetics, in treatment of arthritis, spasm, hair loss, digestive complications and rheumatism.

Cynodon dactylon (Doob grass)

It is a good fodder grass; the plant sap is used in menstruation, diabetes, constipation, indigestion, piles, eye problems and skin diseases.

Desmostachya bipinnata (Daab/Kush)

Stems are used to make brooms, leaves are used in thatching, making ropes and cordage, root decoction is taken orally in diarrhea, leucorrhea and piles.

Hordeum vulgare (Barley/ Jau)

Dry stem after harvesting used as straw and mulch. Seeds are digestive used to cure fever, stomachache and lower the blood sugar. It is also an important gradient of *Sattu* a local dish that is coolant, digestive rich in fiber, calcium, iron etc.

Phragmitis karka (Narkul/ tall reed)

Young shoots are cooked as vegetable, panicles are used as broom, stem is used for making mats and houses. Root is effectively used in urinary troules, heart disease and as aphrodisiac.

Liliaceae

Aloe vera (Alovera/Ghritkumari)

The *Aloevera* gel is used in wound healing, skin disease, diabetes, ulcer, cancer and in moisturiser and cosmetics.

Asparagus racemosus (Satavari)

Roots are used in the form of paste, powder and decoction in treatment of skin disease, jaundice, diabetes, and indigestion. Stomachache, dysentery, urinary disorders, bleeding in urine and in sexual problems.

Allium cepa (Onion/Pyaz)

The bulbs are used as vegetable, is used as anti-diabetic and in hypertension.

Gloriosa superva (Kalihari)

Leaves and root are used to treat ulcer, piles,asthma, urinary disorders, gonorrhea, snakebite and in smooth child birth. Seeds are also medicinally important used in treatment of various diseases.

Smilax ovalifolia (Kumarika/ Bhitura)

Tender leaves are edible; leaf sap is used in treatment of skin diseases. Stem bark is used an anti-malarial. Roots are used in jaundice, rheumatism, dysentery, arthritis, sexual diseases and wound healing.

Santalaceae

The plants of this family are herbs, shrubs and trees, mostly semi-parasite with simple, entire and opposite leaves. Inflorescence is raceme, spike, head or dichasial and achine or drupe type fruit.

Santalum album (Chandan)

The wood is used in funeral of dead body, used in marking *Tialk*, carving and handicraft. Sandal wood oil obtained from heart wood used in perfumery and cosmetic industry. It is very useful in treatment of headache, fever, gonorrhoea, skin disease and in urinary problems, seeds are abortifacient (Jain, 1991).

Osyris lanceolata (African sandalwood)

Wood is very hard and strong, used in carving and building poles. Tannin is obtained from stem bark. Roots are excellent source of red dye, root fibers used in making basket. The twigs are used for making brooms, wood oil is used in perfumery, stem and root bark, leaves are use to cure anaemia, headache, diarrhea, dysentery and making tonic (Orwa *et al.*, 2009).

Thesium wightianum (Himalayan Toadflax)

The whole plant is tied on cheeks to prevent bulging (Murugesan *et al.*, 2005)

Pinaceae

Pinaceae family comes under gymnosperm group of plants. Plants are dominantly trees and few shrubs. These are also called softwood species, resin canals in wood, bears needle shaped leaves, large female cones, two winged seeds, pollination by wind.

Abies pindrow (Himalayan silver fir)

It has beautiful architecture therefore used as outdoor ornamental plant. Wood is used as fuel, construction material and small tools. The bark and leaves are used in treatment of cough, bronchitis, asthma and in pulmonary complications.

Cedrus deodara (Deodar)

The wood of deodar is durable, rot resistant used in construction of temples, houseboats in Srinagar. Wood is also used in making storage bins for maize, paddy etc. Grains are safe from insects and pests due to cedar oil present in deodar wood. The cedar oil has an antifungal and insect repellent property therefore it is used locally as spray against ticks and mites on the cattle.

Pinus roxburghii (Chir/Chir pine)

Wood is moderate in strength and used for light construction works and for decorative furniture. Seeds are edible after cook or uncooked. Resin obtained from stem and leaves are used as antiseptic, diuretic, vermifuse, cough, cold in snakebite, burns, skin complications etc.

Picea smithiana (Himalayan Spruce)

Wood is very light therefore, chiefly used for construction of packing cases, match boxes and fuel purpose. Resin is used in treatment of cardiovascular problems.

Thymelaeaceae

Herbs and vines are very few, tree and shrubs are dominant with tough and fibrous bark Inflorescence generally racemose some time capitate. Calyx tubular and caducous. Phloem of the members of this family contains very strong fibers that makes bark suitable for paper industry.

Aquilaria agallocha (Agar/ Eaglewood)

The wood used for carving, making match boxes and veneer. The fragrance by burning agarwood valued in various rituals in Buddhism. The resin obtained from stem used in production of incense and perfumes, soaps and shampoo. The leaves are used in making tea which is improves constipation.Traditionally plant sap and bark is used to cure ulcer, fever, edema, jaundice, diarrhoea, rheumatism, vomiting, gout and skin complaints.

Daphne oleoides (Agru/Kalishadi)

Wood is used as fuel. Fruit is edible, useful in rheumatism and pimples. Root and leaves are used to cure gonorrhea, indigestion problems and skin diseases.

Daphne bholua (Agru/Kalishadi)

Stem and root yields fibers, which are used in making ropes and paper respectively. Stem bark is used in treatment of fever. Root is used to cure indigestion complaints and as a vermifuse. Plant is also used as stimulant to buffalos for mating (Malla and Chhetri, 2009).

Edgeworthia tomentosa (Oriental Paper Bush)

It is a beautiful shrub used as an ornamental. Bark is used in fish poisoning, bark fibers are used for handmade paper.

Plates of Wild Ethnomedicinal Plants of India

Plate I: (A) *Anticharis senegalensis* (Walp.) Bhandari., **(B)** *Barleria acanthoides* Vahl., **(C)** *Bergiaam manoides* Roth., **(D)** *Blumea obliqua* (L.) Druce, **(E)** *Boerhavia repens* L., **(F)** *Borreria articularis* (L.f) F.N.. Willams.

Plate II.(A) *Bouchea marubifolia* (Fenzl ex Walp.) Schau., **(B)** *Cassia angustifolia* Vahl., **(C)** *Cassia pumila* Lam., **(D)** *Celosia argentea* L., **(E)** *Cleome gracilis* Edgew., **(F)** *Cleome gynandra* L., **(G)** *Cleome vahliana* Farsen., **(H)** *Convolvulus microphyllus* Sieb. ex Spreng.

Plate III (A) *Corbichonia decumbens* (Forsk.) Excell, **(B)** *Corchorus depressus* (L.) Stocks, **(C)** *Crotalaria medicaginea* Lam., **(D)** *Cyperus pygmaeus*Rottb., **(E)** *Datura metel* L., **(F)**, **(G)** *Dipcadi erythraeum* Webbl. & Berth.

Plate IV (A) *Dipteracanthus patulus* (Jacq.) Nees, **(B)** *Eclipta alba* (L.) Hassk., **(C)** *Euphorbia chamaesyce* L., **(D)** *Euphorbia granulate* Forssk., **(E)** *Fagonia schweinfurthii* (Hadidi) Hadidi, **(F)** *Gisekia pharnaceoides* L.

Plate V(A) *Glinus lotoides* L., **(B)** *Goniogyna hirta* (Willd.) DC, **(C)** *Heliotropium strigosum* willld. **(D)** *Heliotropium ovalifolium* Forssk., **(E)** *Indigofera hochstetteri* Baker, **(F)** *Indigofera linifolia* (L.f.) Retz.

Plate VI (A) *Indigofera linnaei* Ali., **(B)** *Justicia procumbens* L., **(C)** *Kohautia aspera* (Heyne ex Roth) Brem., **(D)** *Launaea procumbens* (Roxburgh) Ramayya & Rajagopal, **(E)** *Launaea resedifolia* (L.) Kuntze, **(F)** *Lepidagathis trinervis* Nees.

Plate VII (A) *Leucas cephalotes* (Koen. ex Roth) Spreng, **(B)** *Limeum indicum* Stocks. ex T. Anders, **(C)** *Lindenbergia urticifolia* Lehm., **(D)** *Mollugo cerviana* (L.) Seringe, **(E)** *Mollugo nudicaulis* Lam., **(F)** *Phyllanthus maderaspantensis* L.

Plate VIII (A) *Polycarpaea corymbosa* (L.) Lam., **(B)** *Psoralea plicata* Delile: Plant, **(C)** *Psoralea plicata* Delile: Branch, **(D)** *Pulicaria crispa* (Forssk.) Benth. & Hook. f., **(E)** *Pulicaria wightiana* DC Clarke., **(F)** *Salvia santolinaefolia* Boiss.

Plate IX (A) *Seetzenia lanata* (Willd.) Bullock, **(B)** *Sesuvium portulacastrum* L., **(C)** *Solanum albicaule* Kotschy ex Dunal, **(D)** *Solanum xanthocarpum* Schrad. & Wendl., **(E)** *Tephrosia leptostachya* DC., **(F)** *Tephrosia purpurea* (L.) Pers.

Plate X *Tephrosia strigosa* (Dalz.) Sant. & Maheshwari, **(B)** *Tribulus pentandrus* Forsk var. *pentandrus*, **(C)** *Tribulus rajasthanensis* Bhandari &. Sharma, **(D)** *Trichodesma indicum* (L.) R.Br., **(E)** *Verbesina encelioides* (Cav.) Benth. & Hook. f., **(F)** *Vigna trilobata* (L.) Verd.

Plate XI (A) *Cadaba fruticosa* (L.) Druce, **(B)** *C. fruticosa*: flower, **(C)** *Cocculus pendulus* (J.R.Forst. &G. Forst.) Diels, **(D)** *Dactyliandra welwitschii* Hook. f., **(E)** *Ephedra foliata* Boiss. & Kotschy ex Boiss., **(F)** *Merremia aegyptia* (L.) Urb.

Plate XII (A) *Maerua oblongifolia* (Forssk.) A. Rich: flower, **(B)** *M. oblongifolia*: fruit, **(C)** *Pentatropis spiralis* (Forssk.) Decne., **(D)** *Pergularia daemia* (Forsk.) Chiov., **(E)** *Rhynchosia schimperi* (Hochst. & Steud. ex Steud.) Boiss., **(F)** *Rivea hypocrateriformis* (Desr.) Choisy

Plate XIII (A) *Capparis decidua* (Forssk.) Edgew **(B)** *Aerva persica* (Burm. f.) Merr., **(C)** *Aerva pseudotomentosa* Blatter & Hallberg., **(D)** *Calligonum polygonoides* L., **(E)** *Acacia jacquemontii* Benth., **(F)** *Mimosa hamata* Willd.

Plate XIV (A) *Crotalaria burhia* Buch.-Ham., **(B)** *Euphorbia caducifolia* Haines., **(C)** *Grewia tenax* (Forssk.) Fiori, **(D)** *Leptadenia pyrotechnica* (Forsk.) Decne, **(E)** *Lycium barbarum* L., **(F)** *Vitex negundo* L.

Plate XV (A) *Epilobium latifolium* **(B)** *Cyananthus lobatus* **(C)** *Dactylorhiza hatagirea* **(D)** *Inula grandiflora*, **(E)** *Taxus wallachiana* **(F)** *Allium wallichii.*

Plate XVI (A) *Berberis aristata* **(B)** *Bistorta affinis* **(C)** *Colquhounia coccunea* **(D)** *Corydalis meifolia* **(E)** *Elsholtzia eriostachya* **(F)** *Gaultheria trichophylla.*

Plate XVII (A) *Nephenthes khasiana* **(B)** *Vetiveria zizanioides* **(C)** *Altingia excelsa,* **(D)** *Azalea indica,* **(E)** *Mesua ferrea* **(F)** *Coleogyne nitida.*

Plate XVIII (A) *Mimusops elangi* **(B)** *Oxalis Debilis,* **(C)** *Chlorophytum tuberosum,* **(D)** *Chlorophytum borivilianum* **(E)** *Oenothera rosea* **(F)** *Elaeocarpus ganitrus.*

Plate XIX (A) *Alangium salvifolium* **(B)** *Asparagus racemosus,* **(C)** *Tylophora asthmatica,* **(D)** *Commiphora wightii*, **(E)** *Andrographis paniculata*, **(F)** *Curcuma amada*.

Plate XX (A) *Phyllanthus acidus,* **(B)** *Rauvolfia serpentina,* **(C)** *Ixora pavetta,* **(D)** *Bridelia retusa,* **(E)** *Buchanania lanzan,* **(F)** *Butea monosperma.*

Plate XXI (A) *Piper nigrum,* **(B)** *Mucunia puriens,* **(C)** *Bombex ceiba,* **(D)** *Stereospermum chelonoides*, **(E)** *Tecomella undullata*, **(F)** *Dichrostachys cinerea.*

Appendices

APPENDIX-1: Selected Ethno-Medicinal Plants and their Utility: At a Glance

	Botanical Name	Common name	Family	Part(s) used	Uses
1.	*Rauvolfia serpentina*	Sarpagandha	Apocynaceae	Tuberous roots contain alkaloid reserpine	Use to lower blood pressure, cure mental disorder, diarrhea nd dysentary
2.	*Withania sominifera*	Ashwagandha	Solanaceae	Roots conatins alkaloids withananine, sonmiferin, somniferinine, nicotine etc.	Use in rheumatism, and in cough & ulcer treatement
3.	*Aconitum nepellus*	Aconite	Ranunculaceae	Tuberous roots conatins alkaloid aconite	Use in rheumatism and neuralgia
4.	*Panax ginseng*	Ginseng	Araliaceae	Roots contain Panaquilon glycoside	Use as stimulant and aphrodiasic.
5.	*Ephedra sinica*	Ephedra	Gnetaceae	Green stem & leaves contain alkaloid ephedrine	Use in asthma
6.	*Chinchona officinalis*	Quinine	Rubiaceae	Bark	Use in treatment of malaria
7.	*Strychnos nux-vomica*	Nux-vomica	Loganiaceae	Seeds contain alkaloids like strychnine and brucine	Use in nervous disorders and paralysis
8.	*Papaver somiferum*	Opium	Papaveraceae	Latex from unripe fruts contains alkaloids like morphine, codeine, narceine, papaverine etc.	Alkaloi morphine relievs pain and codeine is analgesic
9.	*Astemisia cina*	Santonin	Asterace	Flower buds	Use as purgative as weel as anthelminthic
10.	*Atropa belladonna*	Belladona	Solanaceae	Succulent leaves	Use in gonorrhea, skin diseases and disease of liver and spleen.

11. *Datura stramonium*	Dhatura	Solanaceae	Leaves contain alkaloid Atropine, hyoscyamine, scopolamine, solanine etc.	Use in relaxing bronchial muscles nad treating asthma
12. *Erythroxylum coca*	Cocaine	Erythroxyla-ceae	Leaves contain alkaloid cocaine	Use as local anaesthesia
13. *Croton tiglium*	Jamalghota	Euphorbiaceae	Seed oil	Use as powerful purgative
15. *Catharanthus roseus*	Periwinkle	Apocynaceae	Leaves and flower buds contain alkaloids vinblastin and vincrystine	Use to cure diabetes

APPENDIX-2: Important Dye Yielding Plants of India

S. No.	Botanical Name	Common name	Family	Part(s) used	Dye Obtained
1	*Acacia catechu*	Khair	Mimosaceae	Wood	Red and brown dye
2	*A. farnesiana*	Gand babul	Mimosaceae	Bark	Black
3	*A. leucophloea*	Ronz	Mimosaceae	Bark	Red
4	*Aegle marmelos*	Bel	Rutaceae	Fruit	Yellow
5	*Annona reticulata*	Ramphal	Annonaceae	Fruit	Red
6	*Anogeissus latifolia*	Kalghi, safed dhok	Combertaceae	Leaf	Black
7	*Artocarpus lakoocha*	Dahua	Moraceae	Root	Yellow
8	*A. heterophyllus*	Kathal	Moraceae	Wood	Yellow
9	*Bauhinia purpurea*	Kachnar	Caesalpiniaceae	Bark	Purple
10	*Bixa orellana*	Annatto	Bixaceae	Aril	Red-orange
11	*Butea monosperma*	Tesu or Flame of Forest	Fabaceae	Flowers	Organge-yellow
12	*B. superba*	Palas lata	Fabaceae	Flower	Yellowis-Orange
13	*Caesalpinia sappan*	Sappan	Caesalpiniaceae	Wood	Red
14	*Cassia auriculata*	Anwal	Caesalpiniaceae	Flowers	Yellow
15	*Cassia tora*	Puad		Seed	Blue or red
16	*Carthamus tinctorius*	Safflower	Asteraceae	Flower	Yellow
17	*Cedrela toona*	Toon	Meliaceae	Leaf	Black
18	*Crocus sativus*	kesar	Iridaceae	Style and stigma	Yellow
19	*Curcuma longa*	Turmeric	Zingiberaceae	Rhizome	Yellow
20	*Diospros peregrine*	Gab	Ebnaceae	Fruit	Brown
21	*Emblica officinalis*	Aonla	Euphorbiaceae	Bark	Brown
22	*Garuga pinnata*	Chogar		Leaf	Red
23	*Tectona grandis*	Teak	Verbenaceae	Bark	Yellow
24	*Indigofera tinctoria*	Indgo or neel	Papilionaceae	Leaves	Blue
25	*Lawsonia inermis*	Henna or mehndi	Lythraceae	Leaves	Orange
26	*Mallotus philippinensis*	Kamla, sindoor	Euphorbiaceae	Fruits	Red
26	*Morinda tinctoria*	Noni	Rubiaceae	Bark, root	Red
27	*Nyctanthes arortristis*	Harsingar	Nyctanginaceae	Flower	Yellow
28	*Onosma hispida*	Ratanjot	Boraginaceae	Roots	Red
29	*Pinus roxburghii*	Chir pine	Pinaceae	Bark	Yellow
30	*Pterocarpus santalinus*	Raktchanda	Papilionaceae	Wood	Red
31	*Rhamnus globosa*	Lakao	Rhamnaceae	Bark	Green
32	*Shorea robust*	Sal	Dipterocarpa-ceae	Bark	Red & black
33	*Terminalia bellirica*	Bahera	Combertaceae	Fruit	Yellow
34	*Woodfordia fruticosa*	Dhawai	Lythraceae	Flower	Brick red
35	*Ziziphus mauritiana*	Ber	Rhamnaceae	Leaf	Red-pink

APPENDIX-3: List of various uses of wood by tribes of India

S.No.	Botanical Name	Common name	Ethnobotanical use
1.	*Butea monosperma*	Chhola	Stem or branch wood is used to make *Khammal Toran* (Marriage pillar) dugout inside the marriage mandap and worship during marriage ceremony.
2.	*Lantana camara*	Kutri/Lantan	Stem/branches are used as fencing around their houses.
3.	*Lagerstromia parviflora*	Senji	Stem wood is used in making frames for thatching, *harsh* for making plough, leaves are used for thatching.
4.	*Azadirachta indica*	Neem	Wood is used for making cots, furniture and as fuel; very thin twigs are used as toothbrush.
5.	*Madhuca latifolia*	Mahua	Wood is used as timber and in making frames of thatching, *Mathani* used for separation of butter from the curd and as a fuel wood.
6.	*Anogeissus pendula*	Dhava	Excellent fuel wood, used for making agriculture implements.
7.	*Acacia nilotica*	Desi kikar	Extremely hard, used as timber, twigs are also used in fencing around agriculture fields.
8.	*Calotropis gigentea*	Akaua	Stem wood is used as fuel and roots are used for making toys.
9.	*Mangifera indica*	Ama	Wood is light and used for making furniture and planks for doors.
10.	*Jetropha gossypium*	Andi	Whole plant is used as fuel wood. Stem is also used for getting stem fibers.
11.	*Terminalia arjuna*	Kahwa	Wood of Arjun is water durable and used for making walls of wells and medicinally.
12.	*Feronia limonia*	Kaitha	Wood is used for making agriculture impliments, Takhats (cot) and as a fuel wood.
13	*Acacia catechu*	Khairi	Agriculture implemnts
14	*Lannea coromandelica*	Jhingan	Wood is used for making furniture, thatching frames and as fuel wood
15	*Elaeodendron glaucum*	Mamri	Wood is soft and used as fuel wood
16	*Ficus religiosa*	Pipal	Wood is used as fuel wood for cooking food/ prasad in religious ceremonies like bhagawat, katha etc. and other rituals like havan.
17	*Ougenia oojeneinsis*	Sandan	Wood is very hard used for making furniture, agriculture implements and small branches as a fuelwood.
18	*Zizyphus mauritiana*	Beri	Wood is used for making foots of cot and other small furniture, like *Mathani* used for separation of butter from the curd and as a fuel wood.

19	*Cordia dichotoma*	Lamer	Wood is used as timber for making door frames, *Takhats* (cot) and small branches are used as fuel wood.
20	*Diospyros melanoxylon*	Tendu	Wood is light and closely grained used in making furniture, house construction and for agriculture tools.
21	*Mitragyna parvifolia*	Kaima	Wood is not very hard and used for making agriculture tools, as fuel wood in funerlas and in various ritauals
22	*Tectona grandis*	Sagon	Very important timber used forconstruction of houses, beams and furniture.
23	*Morinda citrifolia*	Noni	Wood is used for making frames of thatches, pole and as a fuel wood.
24	*Holoptelia*	chamrel	Wood is moderately hard and used for furniture and as a fuelwood.
25	*Syzygium cumini*	Jamun	Wood is durable in water therefore used inside the wells for making walls and for making wheels of bullock cart.

APPENDIX-4: Examinatioon Tool: Model Question Papers

Question Paper-1

Time: 2.00 hrs **MM: 50**

VERY SHORT ANSWERS TYPE QUESTIONS **(10 marks)**

Q. 1 Match the following list (A) with list (B):

S.No	Plants (A)	Family (B)
1	*Aegle marmelos*	Pinaceae
2	*Saraca asoca*	Euphorbiaceae
3	*Anthocephalus kadamba*	Meliaceae
4	*Azadirachta indica*	Rubiaceae
5	*Terminalia chebula*	Comberetaceae
6	*Diospyros melanoxylon*	Mimosaceae
7	*Acacia nilotica*	Ebnaceae
8	*Emblica officinalis*	Fabaceae
9	*Dendrocalamus strictus*	Rutaceae
10	*Picea smithiana*	Poaceae

Q.2 Fill in the blanks

(a) .... coined the term ethnobotany.

(b) Bark of Cinchona officinalis used as ....

(c) .... known as father of Indian Ethnobotany.

(d) Botanical name of Bel is ....

SHORT ANSWER TYPE QUESTIONS **(20 marks)**

Q.3 Attempt any five

1. Define following words
   Carminative, Anodynes, Expectorant, Abortifacient.
2. Write a short note on Ethnoecology?
3. Write a short note on Ethnonarcotics.
4. Write short note on mythology of Ashoka Tree.
5. Write down the ethnobotanical uses of the following plants
   Cynodon dactylon, Barleria cristata, Ageratum conyzoides, Ocimum sanctum
6. Write down the mythologyof Tulsi.

LONG ANSWER TYPE QUESTIONS

Give answer of any two questions **(20 marks)**

Q.4 Describe indigenous culture. Give mythology of mango and Sandalwood tree?

Q.5 Discuss the importance and scope of ethnobotany.

Q.6 Discuss ten dye-yielding plants along with their scientific names, families and folk uses.

Q.7 Give ethnomedicinal uses of the plants of family apocynaceae.

Question Paper-2

VERY SHORT ANSWERS TYPE QUESTIONS **(10 marks)**

Q.1 Fill in the blanks

1. Study of culture .... study of plant ....
2. Curcumin is obtained from .... part of the .... plant.
3. Gum guggal is obtained from ....
4. CIMAP stand for .... and located at ....
5. .... is an alkaloid derived from the bark of the .... tree use in malaria disease.
6. Botanical name of Safed musli is ....
7. Hingis obtained from ....
8. Cinnamomis obtained from ....
9. .... is called is the father of ethanobotany in India.
10. Bhil and meena are tribes of .... state.

Q.2 Multiple Choice Questions

1. The state with maximum percentage of tribal population is ....
   a) Uttar Pradesh b) Punjab
   c) Madhya Pradesh d) Rajasthan
2. Botanical name of Arjun tree
   a) *Terminalia arjuna* b) *Terminalia malabaricum*
   c) *Tectonagrandis* d) *Lagerstroemia parviflora*
3. First work started on ethanobotany in India
   a) JanikiAmmal of BSI b) BirbalSahni of NBRI
   c) S.K.Jain of BSI d) D. Brandis of FRI
4. An "Introduction to Ethanobotany" is written by:
   a) S.K. Jain b) Harsberger
   c) Faulkes d) Janiki Ammal

SHORT ANSWER TYPE QUESTIONS **(20 marks)**

Q.3 Write short note on any five of the following

a) Ethanoforestry b) Sacred Plants
c) *Withania somnifera* d) Medicinal uses of Amla
e) Mythology of *Varuna* f) Lepcha tribe

LONG ANSWER TYPE QUESTIONS (20 marks)

Give answer of any two questions

Q.4 Name the Ethanobotanical plants from the family of Rosaceae.

Q.5 Write a note on botanical gardens.

Q.6 Tribals are protector or destroyer of forest wealth, explain it?

Q.7 Write a detail note on the role of ethanobotany in conservation of plant genetic resources.

Question Paper-3

VERY SHORT ANSWERS TYPE QUESTIONS (10 marks)

Q.1 Fill in the blanks

1. Botanical name of Rudraksha is ....
2. The term ethanobotany was first coined by ....
3. "Introduction of Ethanobotany" is written by ....
4. Amla is rich source of ....
5. Guggal belongs to the family ....
6. Indigenous medicine is also known as ....
7. The wood of .... is used for sandal wood oil.
8. Weeping willow belongs to the family ....
9. Castor oil is obtained from ....
10. The study of how people of a particular culture and region make of use of indigenous plants is known as ....

Q.2 Multiple Choice Questions

1. Beuteamonosperma belongs to family:
   a) Fabaceae b) Poaceae
   c) Malvaceae d) Verbenaceae
2. The book " Glimpses of Indian Ethnobotany" is writtenby:
   a) S.K. Jain & A.K. Jain b) V. Shilpa
   c) S.K. Jain d) S.K. Jain and V. Mudgal
3. Wheat and rice belong to the family:
   a) Poaceae b) Rutaceae
   c) Asterecaceae d) Guttiferaceae
4. Botanical name of Sarpgandha
   a) *Withaniasomnifera* b) *Rauvolfiaserpentina*
   c) *Mucunapuriens* d) *Aloe vera*

SHORT ANSWER TYPE QUESTIONS (20 marks)

Q.3 Write short note on any five of the followinng

a) Folktales b) Indigenous knowledge
c) Ethnonarcotics d) Garo tribe
e) Bhujni Batiyani ritual
f) Rapid ethnobotanical appraisal (REA)

LONG ANSWER TYPE QUESTIONS **(20 marks)**

Give answer of any two questions

Q.4 Write the ethanobotanical uses of family Malvaceae, Fabaceae and Lauraceae with suitable examples.

Q.5 Write the definition and the scope of ethanobotany.

Q.6 Tribals are protector or destroyer of forest wealth, explain it?

Q.7 Write the scientific name with family of ten ethanobotanical plants and their uses by tribes inhabiting in Uttrakhand state.

Question Paper-4

VERY SHORT ANSWERS TYPE QUESTIONS (10 marks)

Q.1 Fill in the blanks

1. *Azadirachta indica* belongs to family ....
2. *Datura fastuos a* is used for ....
3. Khasi tribe belongs to .... state of India.
4. Scientific name of rubber yielding plant ....
5. Tharu tribe belong to .... state of India.
6. Akas bail is used for hair treatment. Scientific name of Akas Bail.
7. Grasses belong to family ....
8. Botanical name of finger millet is ....
9. Basil belongs to the genus ....
10. .... is known as father of medicine.

Q.2 Multiple Choice Questions

1. Which among these tribes are not found in Rajasthan:
   a) Bhil b) Damar
   c) Dhanka d) Kani
2. Lotus is a symbol of:
   a) Goddess Luxmi b) Lord Shiva
   c) Lord Krishna d) Goddess Parvati
3. The state with maximum percentage of tribal population is...........
   a) Uttar Pradesh b) Punjab
   c) Madhya Pradesh d) Rajasthan
4. Botanical name of Arjun tree
   a) *Terminalia arjuna* b) *Terminalia malabaricum*
   c) *Tectona grandis* d) *Dipterocarpus arjuna*

SHORT ANSWER TYPE QUESTIONS (20 marks)

Q.3 Write short note on any five of the following

a) Indigenous culture b) Ethanotoxicology

c) Myhology of Bael d) Totem

e) Taboo f) Folklores

LONG ANSWER TYPE QUESTIONS **(20 marks)**

Give answer of any three questions

Q.4 Name two important institution involve in ethanobotany with their contribution.

Q .5 Write the botanical name, Family, part used and medicinal uses of important five ethanobotanical plants.

Q .6 Write in detail the contribution of ethenic or tribal groups in enhancing the knowledge of ethanobotany in India?

Q.7 Define Magico-religious therapies of Monpas tribe in detail.

Question Paper-5

VERY SHORT ANSWERS TYPE QUESTIONS (10 marks)

Q.1 Fill in the blanks

1. Botanical name of "Salam Panja" is ....
2. Khombo tribe belongs to .... state of India.
3. Rosa indica belongs to the family ....
4. .... is the study of the traditional medicine practiced by various ethnic groups.
5. Khasi and Garo tribes inhabiting in .... state of India.
6. Castor belong to family is ....

Q.2 Multiple Choice Questions

1. First work started on ethanobotany in India
   a) Janiki Ammal of BSI b) Birbal Sahni of NBRI
   c) S.K.Jain of BSI d) D. Brandis of FRI
2. An "Introduction to Ethanobotany" is written by:
   a) S.K.Jain b) Harsberger
   c) Faulkes d) Janiki Ammal
3. Beutea monosperma belongs to family:
   a) Fabaceae b) Poaceae
   c) Malvaceae d) Verbenaceae
4. AICRPE (All India Co-ordinated Research Project on Ethanobotany) started by:
   a) MoEF (1992-1998) b) DST (1992-2000)
   c) MoAG (2000-2004) d) None of the above
5. Cotton belongs to the family:
   a) Poaceae b) Rutaceae
   c) Malvaceae d) Guttiferaceae
6. Botanical name of *Barna* tree
   a) *Withania somnifera* b) *Hopea odorata*
   c) *Crataeva nurvala* d) *Bridelia retusa*
7. Which among these tribes are not found in Rajasthan:
   a) Bhil b) Damar
   c) Dhanka d) Kani
8. Barna is associated with:
   a) Goddess Luxmi b) Lord Shiva
   c) Lord Krishna d) Goddess Parvati

9. Coffee belongs to family:
   a) Poaceae b) Rutaceae
   c) Rubiaceae d) Malvaceae

SHORT ANSWER TYPE QUESTIONS (20 marks)

Q.3 Write short note on any five of the following
   a) Ethanohorticulture b) Conservation
   c) Gond Tribe d) *Aloevera*
   e) Decoction f) Ethanomedicine

LONG ANSWER TYPE QUESTIONS (20 marks)

Give answer of any three questions

Q.4 Write in detailed notes on ethanoforestry and sacred grooves with suitable example.

Q.5 Define the term Ethanobotany, what is the relation between tribal community and forest?

Q.6 Write folk lores associated with following plants: Peepal, Bargad, Madhuca.

Q.7 Explain the traditional knowledge of ethanomedicine. Write botanical name and family of five plant used for ethanoveterinary purposes.

Bibliography

Alcorn JB (1984). Haustec Mayan Ethnobotany, University of Texas Press, Texas.

Alexiades MN (1996). Collecting ethnobotanical data: An introduction to basic concepts and techniques. In: Alexiades M.N. (ed) Selected Guidelines for Ethnobotanical Research: A Field Manual. The New York Botanical Garden, Bronx, New York, pp. 53-94.

Anderson EN (2001). Comments. In: Ford, R., editor. Ethnobiology at the millennium: past promise and future prospects. Ann Arbor (MI): Museum of Anthropology, 175–186.

Anon (1998). Status report on biodiversity. Ministry of Enviroment and Forest. GOI, New Delhi

Arora RK (1991). Conservation and Management concept and Approach in Plant Genetic resources. (eds.) Paroda R.S.& R.K.Arora, IBPGR, Regional office South and Southeast Asia, New Delhi, pp. 25.

Baishya, D (2009). Traditional healthcare practices of Monpas Tawang, Arunachal Pradesh. Ph.D. Thesis, Pp. 262.

Balick MJ and Paul AC (1996). Plants, People, and Culture: The Science of Ethnobotany. New York: Scientific American Library.

Berlin B (1972). Speculations of the growth of ethnobotanical nomenclature. *Lang Soc*, 1: 51–86.

Berlin B, Breedlove D, Raven P (1966). Folk taxonomies and biological classification. *Science*, 154: 275.

Bharti AK and Sharma BL (2012). Plants used as Ethnoveterinary Medicines in Sikkim Himalayas, *Ethnobotnay Resaerch & Application, 10: 339-356.*

Bohra N, Tewari LM, Tewari A (2017). Ethnobotany of wild edible plants traditionally used by the local people in the Ramnagar region from Nainital District, Uttarakhand, India. *Biolife*, 5(1):12-19.

Brown C (1984). Language and living things: uniformities in folk classification and naming. New Brunswick (NJ): Rutgers University Press.

Carlson T and Maffi L (2004). Ethnobotany and conservation of biocultural diversity. Advances in economic botany. 15. New York: Botanical Garden Press.

Chakrabart F, Kisku AK and Dolai MC (2012). Health maintaining and disease curative ethno-medicinal and religious practices by the Santals of Keonjhar district, Orissa. *Journal of Humanities and Social Sciences*, 2(5): 35-45.

Chakraborty T and Vasudeva Rao MK (1988). Ethnobotanical studies of the shompen of Great Nicobar Islands, *J Econ Taxon Bot,* 12: 43-54.

Conklin H (1954). The relation of Hanuno′o culture to the plant world (PhD Dissertation). New Haven (CT): Yale University.

Cunningham AB (2001). Applied Ethnobotany. People, Wild Plant Use and Conservation. WWF, UNESCO, Royal Botanical Gardens, Kew. Earthscan Publications, London and Sterling.

Dagar H S (1989). Plant folk medicines among Nicobarese tribal of Car Nicobar Island, India, *Econ Bot,* 43: 215-224.

Das S, Sheeja TE and Mandal AB (2006). Ethnomedicinal uses of certain plants from Bay Islands, *IJTK*, 5(2): 207-211.

Davison-Hunt I (2000). Ecological Ethnobotany: stumbling towards new practices and paradigms. In: *MASA Journal*, 16 (1): 1-13.

Davis W (2001). Light at the edge of the world: a journey through the realm of vanishing cultures. Vancouver (BC) and Washington (DC): Douglas & McIntyre Press, and National Geographic Society.

Deb RS and Sharma BK (2015). Traditional Healing Practices in North East India, *IJHS*, 50(2): 324-332.

Dey A (2015). Plumeria rubra L. (Apocynaceae): Ethnobotany, phytochemistry and pharmecology: a mini review. *Journal of Plant Sciences*, 10(2): 54-62.

Deur D, Turner NJ (2005). Keeping it living: traditions of plant use and cultivation on the Northwest coast of North America. Seattle (WA) and Vancouver (BC): University of Washington Press, and UBC Press.

Dougherty J (1978). Salience and relativity in classification. *Am Ethnol*, 5(1):66–80.

Dutt HC, Bhagat N and Pandita S (2015). Oral traditional knowledge on medicinal plants in jeopardy among Gaddi shepherds in hills of northwestern Himalaya J&K India. *Journal of Ethnopharmacology*,168: 337-48.

Dutta BK, Dutta PK (2005). Potential of ethnobotanical studies in North East India: an overview. *Indian Journal of Traditional Knowledge*, 4:7–14.

Ertug F (2006). Ethnobotany: at the junction of the continents and disciplines. Conference proceedings of the Fourth International Congress of Ethnobotany (ICEB 2005). Istanbul, Turkey: Yeditepe University, Yayinlari; 2006.

Faulks PJ (1958). An introduction to Ethnobotany. Moredale, London.

Ford RI (1994). The nature and status of ethnobotany. Anthropological Papers 67. Ann Arbor: Museum of Anthropology, University of Michigan.

Friedman J, Yaniv Z, Dafini A, Palewitch D (1986). A preliminary classification of the healing potential of medicinal plants, based on a rational analysis of an ethnopharmacological field survey among Bedouins in Negev desert, Israel. *Journal of Ethnopharmacol*, 16(2-3): 275-287.

Gadgil M and Vartak VD (1976). Sacred groves of Western ghats in India. *Economic Botany*, 30: 152-160.

Gerique A (2006). An introduction to ethnoecology and ethnobotany Theory and Methods. Tech. Rep. Pp. 20.

Ghosh A (2008). Ethnomedicinal plnats used in West Rarrh region of West Bengal, *Natural Products Raduance*, 7(5): 461-465.

Hamilton AC, Shengji P, Kessy J, Khan AA, Lagos-Witte S and Shinwari ZK (2003). The purposes and teaching of Applied Ethnobotany. People and Plants Working Paper 11. WWF, Godalming, UK

Harshberger JW (1896). The purpose of ethnobotany. *Botanical Gazet.* 21: 146-158.

Jain SK (2001). Ethnobotany in modern India. Phytochemistry, Golden Jublee issue, 39-54.

Jain SK (2002). Bibliography of Indian ethnobotany. India: Scientific Publishers.

Jones, VH (1941). The nature and scope of ethnobotany. *Chronica Botanica*, 6(10): 219-221.

Joseph J and Kharkonger A (1997). A preliminary ethnobotanical survey in Khasi and Jantia Hills, Meghalaya: Contribution to Indian Ethnobotany (ed. Jain SK), Scientific Publishers, Jodhpur, pp 187-194.

Kosambi, DD (1991). The culture and civilisation of ancient India in historical outline. Vikas Publishing House, New Delhi, India.

Kumar R, Singh MK and Mudgal P and Bharti KA (2014). Medicinal and Aromatic Flowering plnats of Uttarakhand: An Overview. *In* Medicinal Plnats: Aspects and Prospects, Biotech Books, New Delhi, Pp: 336-346.

Martin GJ (1995). Ethnobotany: a methods manual. London: Chapman and Hall.

Martin GJ (1995). Ethnobotany. A "People and Plants" Conservation Manual. World Wide Fund for Nature. Chapman & Hall, London.

Minnis PE and Elisensw (2000). Biodiversity and Native North America. Norman: University of Oklahoma Press; 2000.

Nazarea V (1998). Cultural memory and biodiversity. Tucson: University of Arizona Press.

Nazir AP, Negi AK and Todaria N P (2010). Traditional uses of medicinal plants of Pauri Garhwal, Uttarakhand. *Nature and Science,* 8(6): 57-61.

Nolan JM (2007). Wild harvest in the heartland: ethnobotany in Missouri's Little Dixie. Lanham (MD): Rowman and Littlefield.

Nolan M (2002). Wild plant classification in Little Dixie: variation in a regional culture. *J Ecol. Anthropol*, 6(1):69–81.

Orwa C., Mutua. Kindt R., Jamnadass R., and Simons A. (2009). Agroforestry Database: a tree reference and selection guide version 4.0 http://www.world agroforestry.org/ treeb2/ AFTPDFS.

Phillips OL (1996). Some quantitative methods for analysing ethnobotanical knowledge. In: Alexiades M.N. (ed) Selected Guidelines for Ethnobotanical Research: A Field Manual. pp. 171-197. The New York Botanical Garden, Bronx, New York

Plotkin M (1993). Tales of a Shaman's apprentice: an ethnobotanist searches for new medicines in the Amazon Rain Forest. New York: Penguin.

Prasad SK and Singh BN (2014). Ducumentation of Ethno Medicinal Plants of Gopalganj District of Bihar (India), *JPBS,* 9(3): 80-89.

Priyanka BM (2016). An ethno-botanical study of medicinal plants of Ghurdauri region, Uttarakhand. *India Journal of Medicinal Plants Studies,* 4 (5): 200-205.

Pushpagandhan P (1994). Ethnobiology in India. A status report, Ministry of Environment and Forest, GOI, New Delhi.

Rana MS and Samant SS (2011). Diversity, indigenous uses and conservation status of medicinal plants in Manali wildlife sanctuary, North Western Himalaya. *Ind. J. Traditional Knowledge* 10(3): 439-459.

Rao Vasudeva M K (1986). A preliminary report on the angiosperms of the Andaman and Nicobar Islands, *J. Econ. Taxon. Bot.,* 8(1): 107-184.

Rather MA and Baba SA (2015). Traditional Use of Medicinal Plants in Kasmir: A Review, *Resaerch & Reviews: J. of Bio.* 3(4): 26-32.

Roy B (1998). Sociocultural and environmental dimensions of tribal health: A study among the Dimas Kacharis and and the zeme Nagas of North Cachar hills district Assam. Ph.D. thesis, pp.349.

Schultes RE (1962) The role of ethnobotanist in the search of new medicinal plants. *Lloydia*, 25: 257-266.

Schultes RE and Hofmann A (1987). Plants of the gods: origins of hallucinogenic use. New York: Van der Marck Editions.

Sharief MU (2007). Plants folk medicine of *Negrito* tribes of Bay Islands, *IJTK,* 6(3): 468-476.

Shipman AS and Boster JS (2008). Recall, similarity judgment, and identification of trees: a comparison of experts and novices. *Ethos*, 36(2): 171–192.

Singh MK (2017). Distribution and implication of alien invasive plant species of Assam: A review. *Annals of Horticulture* 10 (2): 120-127.

Singh MK, Arya M, Bharti AK and Singh K (2018). Exploration of some folk medicinal herbs in forest fringe villages of Assam (India): A study amid Nagaon and Golaghat districts. *JPP,* 7(1): 2362-2368.

Singh MK, Meena DK, Bharti AK and Kumar R (2017). Study of Forest based Ethno-medicinal plnats among the forest fringe villages of BNP, Meghalaya, *Ann. of Horti.* 10 (2): 128-137.

Sinha SK and Chandra R (2011). Medico-Ethnobotanical Study of Madhubani district Bihar, *Biospectra*, 6 (1): P.P 131-134.

Srivastava R (2014). A review on phytochemical, pharmecological and pharmecognostical profile of Wrightia tinctoria: Adulterant of kurchi. *Pharmecognosy Review*, 8(15): 36-44.

Stepp JR, Castaneda H and Cervone S (2005). Mountains and biocultural diversity. *Mt. Res. Dev.*, 23(3): 223–227.

Tariq A, Mussarat S and Adnan M (2015). Review on ethnomedicinal phytochemical and pharmacological evidence of Himalayan anticancer plants. *Journal of Ethnopharmacology*, 164: 96-119.

Thakur AK and Singh, KJ (2015). Common Medicinal Herbs of Shimla City, Himachal Pradesh, India. In Medicinal Plants: Distribution, Utilization and Significance. Discovery Publishing House Pvt. Ltd., New Delhi, pp: 113-125.

Townsley, P (1996). Rapid Rural Appraisal, Participatory Rural Appraisal and Aquaculture. FAO Fisheries Technical Paper 358, Rome

Turner, N (1995). Ethnobotany today in nortwestern North America. 263-283 in R.E. Schultes and S Von Reis, eds. Ethnobotany: Evolution of discipline. Dioscorides press Portland.

Turnrer N (1996) *Interntional Journal of Ethnobiology and Ethnomedicine.* 2007; 3: 8. Published online 2007 January 30. doi: 10.1186/1746-4269-3-8.

Wickens GE (1990) What is economic botany? *Economic Botany*, 44: 12-28.

Zeven AC and Zhukovsky, PM (1973). Dictionary of cultivated plants and their centers of diversity, PUDOC, Wageningen Pp182.

http//:www.indianmirror.com/ accessed 6 March 2019.

http://www.ecoindia.com/tribes/ accessed 15 Feb 2019.

https://tribal.nic.in/ accessed 19 March 2019.

https://www.quora.com/accessed 19 March 2019.

https://www.tourmyindia.com/blog/the-indigenous-colour-of-india-the-indian-tribes/accessed 19March 2019.

http://www.yourarticlelibrary.com/essay/geographical-distribution-of-tribal-population-in-india/ 41311.

Index

M

N

O

P

FSC
www.fsc.org
MIX
Papier | Fördert gute Waldnutzung
FSC® C083411

Zeitfracht Medien GmbH
Ferdinand-Jühlke-Straße 7
99095 Erfurt, Deutschland
produktsicherheit@kolibri360.de